MW01167590

* * *

Beyond the Weapons of Our Fathers

Edward W. Wood Jr.

Fulcrum Publishing
Golden, Colorado

Library of Congress Cataloging-in-Publication Data
Wood, Edward W., Jr.
Beyond the weapons of our fathers / Edward W. Wood Jr.
p. cm.
Includes bibliographical references.
ISBN 1-55591-179-X (hardcover : alk. paper)
1. Pacifism. 2. Wood, Edward W., Jr. 3. Pacifists—United States—Biography. 4. United States—History, Military. 5. Military weapons—United States. I. Title.
JZ5566.4 .W66 2003
303.6'6—dc21

2002015250

Printed in the United States of America
0 9 8 7 6 5 4 3 2 1

Jacket and book design by Patty Maher

Fulcrum Publishing
16100 Table Mountain Parkway, Suite 300
Golden, Colorado 80403
(800) 992-2908 • (303) 277-1623
www.fulcrum-books.com

This book is dedicated to
Thomas Rauch, surely one of the most compassionate
persons I have ever known,
Mallary Fitzpatrick, my teacher for fifty years,
and to Jack Hall who read everything.

And always . . . Elaine
and My Children

And
To all innocents killed and to be killed
by weapons of mass destruction

And to all those whose works and examples over so many years led me to the making of this book, authors and activists too numerous to list but rooted deep in my heart.

And thanks to Lucy Allen, Jim Aho, Donald Anderson,
Dean Baxter, Gorton Carruth Jr., Sasha Chavchavadze,*
Neil Elgee, Michael Evans-Smith, Stephen Evans, Naomi Horii,
Mary Jones, William Montgomery, Don Mulligan, John McCamant,
Ann Randolph, Bob Reed, Rita Schweitz, Rose Thomson,
David Van Meter, and Nancy Wood
for their assistance and help

And especially to Noa Hall

And always, to Bob Baron
and the staff of Fulcrum Publishing

* *Particularly Gorton Carruth's* American Facts and Dates

✵ ✵ ✵

Contents

What is it that causes violence in America? Why have I turned from guns to a more peaceful way of life? Why am I appalled at bombing runs over Afghanistan, sickened by the killings in the Near East?

Why do I refuse to hate those I am told by my nation to call my enemies?

Why do endless cycles of violence return each generation? What is there about human nature that endlessly repeats cycles of war and peace, violence and compassion?

What is the relationship between our propensity to make enemies and our ever-improving weapons of mass destruction?

What happens when we become the "enemy," attacked by others? When is our response to violence disproportionate to the attack?

Where is the peace I so fervently desire?

Over my lifetime I have struggled with these questions.

This book contains the answers I have discovered.

Part One—

Searching the Past

✯

It is the voices of many forefathers who founded the nation with the weapons of their day that I most often hear as I fall asleep at night, voices of war and peace, violence and compassion ...

CHAPTER ONE

The Weapons of Our Fathers

My Inheritance of Violence
Weapons of My Father and Myself: Two World Wars

My father's .30-06 Winchester rifle introduced me to the world of weapons. I graduated to it from a .22 rifle. A .410 and sixteen gauge shotgun also seduced me into the world of killing as a boy.

I will never remember when Dad bought this weapon. It and all the other rifles and shotguns around our home were common to a southern boy of my generation.

Me with Dad's rifle, 1945.

It introduced me to the fellowship of the weapon: firing it with family and with friends, at targets, then deer; cleaning the powder–dirty barrel, using my thumbnail to reflect light into that darkly mysterious, spiraled chamber.

In World War Two I graduated to the M–1 Garand, the rifle I carried across France in the summer of 1944 during the war against the Nazis.

The M1903 .30 caliber rifle, progenitor of my father's .30–'06, was introduced as the standard infantry weapon for the United States Army after the Spanish–American War of 1898. It and the .45 caliber pistol became effective weapons for protecting and extending the American Empire. This weapon was the standard infantry rifle during World War One and, when the M–1 Garand became the basic weapon in World War Two, was favored by snipers for long-range killing. The Garand, the Browning Automatic Rifle (BAR) and the Light Machine Gun were the mainstays of an infantry platoon in World War Two.

Beginning with King Philip's War in 1675, I had ancestors in every one of America's major wars until 1918, and I served as an infantryman in World War Two.

As a southern boy guns and violence were an essential component of my childhood and early manhood.

I first remember the fog of racial violence in southern towns in Mississippi and Alabama as a boy, Saturday afternoons in small towns where I could feel the violence in the air, as heavy as the humidity that sweated our flesh, as the thunderstorms that brooded over the near horizon. I remember driving down country roads with my father where chain gangs worked the dusty shoulders in black and white striped uniforms, chains on prisoners' ankles glinting in July sun. I remember my father speaking of lynchings on Saturday nights over a bridge table at our home, my mother pursing her lips in disdain at his stories.

I still even own the weapons of my childhood, the weapons my father gave me to introduce me to manhood. The slingshot I used to fire pebbles at unsuspecting robins; the carved wooden pistol I played with, imitating the escape of John Dillinger from a mid-western jail; the Coleman Air Pistol, a far more dangerous weapon, mayhem on birds.

Beginning in the early 1930s, wars haunted the headlines of my childhood. I volunteered for the Army in 1943 at eighteen, the expected duty of a male in my family, then for combat in1944, leaving a safe billet at the University of California at Los Angeles.

From the moment in the battle for the liberation of France in the summer of 1944 when my Division freed slave laborers east of Verdun, from the moment I heard the virulent whisper of artillery shells over my head, the rap of machine gun fire, and shrapnel penetrated my skull, my buttock, my back, I began my search for an understanding of the causes of war and violence and their remedies.

After that war, I struggled with the impacts of my wound, impacts that irrevocably changed my life. For years I imagined an axe protruding from my skull, sunk deep into the place of wounding, pain palpitating in my skull. The pain so intense I sometimes envisioned myself yanking that axe from my brain and using it on others—peers, friends, family, even distant strangers—a throbbing desire to strike back at this world I had never made.

Haunted by memories of the Holocaust, I lived and worked through eras of violence in America: McCarthyism, the War in Korea, Vietnam, ghetto civil disorders, racial bias, our dedication to violence in the media, shocking video games, school shootings, a nation appalled by the tragedies of death and woundings from the newest weaponry. Holocausts in far places such as Cambodia and Rwanda and murderous rampages between Israel and Palestine shadowed our domestic violence. As our wars and battles in third world nations stalked me—Grenada, Panama, Iraq, Somalia, Serbia—they were followed by the horrendous terrorist attack of September 11, 2001, on the World Trade Center, then bombings in Afghanistan, future battles now threatened.

The better part of my adult life has been spent seeking to understand the violence intimate to being an American. Why this capacity to face enemies and seek their destruction? What ability allows us to invent weapons of increasing technological barbarity with each generation, capacities and abilities so seldom admitted in terms of their intensity in the American unconscious and their constancy over American history. From that understanding I have struggled to form an ethic for living with integrity in the midst of this violence, beyond the weapons of our fathers.

That search for meaning, a life beyond violence and without weapons, has involved, most of all, a dialogue with my past: those generations of men who served in all of America's major wars as "citizen soldiers," beginning with the war of the Massachusetts settlers against the Native American Nation of the Wampanoags and their allies in1675. Some of my male forefathers repressed Native Americans, supported slavery (the dark underbelly of my American family) while others fought for the abolition of slavery in the Civil War and dedicated themselves to freedom. Men on both sides in that war were wounded in combat as I was in my war. These were men who discovered their meaning in the wars of their day. Only one of these forefathers, Levi Darwin Tiffany of Blandford, Massachusetts, testified against the violence of his time, questioning the Mexican-American War of 1846.

It is the voices of those many ancestors I hear just before I fall asleep at night—citizen soldiers all. They murmur of the wars they fought, the enemies they defeated and the weapons they carried and so often used. They tell me ours is a nation rooted in war, founded in rebellion and spread over the continent by a ruthless destruction of the Native American. Forged in a Civil War, we catapulted into world power by victory in two great wars, in the Cold War over the Communists and the USSR, and now locked into a vicious struggle with the threat of worldwide terrorism.

A Roll Call of My Forefathers

Citizen Soldiers in the Wars That Formed America Before the Civil War

Humphrey Tiffany 1630–1685

King Philip's War of 1675 in New England. Indian Nations decimated, enslaved, villages burnt. 800 settlers killed. King Philip's head cut off, posted outside Boston.

Consider Tiffany 1733–1796

French and Indian War, 1754–1763, on patrol against the French and Indian enemy in the fall of 1756.

Solomon Lombard 1702–1781

A patriot in the American Revolution, he helped write the Constitution of Massachusetts, served on important Revolutionary War Committees.

Peter Zachary 1745–1797

Also served in the Revolutionary War but in the Caroline County Militia of Virginia. Migrating to Georgia, he homesteaded land in 1787. His descendants helped settle Alabama, take their land from the Creeks, some killed in the 1830s in the warfare between settler and Indian.

James Lombard 1814–1861

Homesteaded land in Texas in 1839.

Stephen Lombard 1819–1848

James' brother, migrated to Mississippi in the 1840s, killed in a lumbering accident on the Mississippi River.

In the Civil War and the World Wars of this Century

William Bigbee Green 1842–1923

Peter Zachary's great grandson, volunteered for the 3rd Alabama Cavalry in September 1861. Wounded at Munfordville in September 1862. Fought for the Confederacy in the west.

Bela Bentley Tiffany 1835–1924

Volunteered for the Union Army in Massachusetts in June 1862. Fought in both west and east, at Knoxville and at Vicksburg. Wounded at Cold Harbor in spring of 1864.

Edward W. Wood 1894–1972

My father volunteered as a pilot for the Air Service in 1918. Cracked up, he walked away and flew Airmail in 1919.

Edward W. Wood Jr. 1924–

The "Last of the Line," volunteered for the Infantry in 1944. Wounded in the liberation of France, September 1944.

And an Early American Pacifist

Levi Darwin Tiffany 1824–1894

Strongly opposed the Mexican-American War of 1846 and did not serve in the Civil War although he was Bela Tiffany's brother.

My Forefathers Tell Me of the Weapons They Carried

Weapons of My American Forefathers
Squire Humphrey Tiffany
FIRE, PIKE AND SNAPHAUNCE
IN KING PHILIP'S WAR, SWANSEA, MASSACHUSETTS: 1675

In 1663 Squire Humphrey Tiffany was permitted to be a "sojourner and to buy and hire" in Swansea, then part of the town of Rehoboth. In 1675 King Philip's War began in Swansea with an attack by the Wamapanoag Indians. That war spread throughout New England, a bitter civil war between conflicting tribes of Indians, the decimation of towns throughout New England. By the end of the war many tribes were enslaved, 50% of settlers' villages burnt or destroyed, and countless Indian women and children murdered along with some white women and children.

Brutality on both sides exceeded imagination, barbarous tortures common, heads posted on poles, Indian homes destroyed. The war followed the patterns of earlier ones in Virginia and Connecticut: the Indian had to be subdued and eradicated if colonists were to succeed.

The Snaphaunce

In the fifteenth century western man first developed a weapon he could hold and fire, a new way of killing quickly imported to the American colonies, able to kill at a distance. Since the Snaphaunce was still extremely primitive by the time of King Philip's War, scarcely accurate, victory over Native Americans also rose from weapons such as the pike and the axe and from fire used to burn out Indian settlements, residents often trapped in the flames, roasted to death.

Weapons of My American Forefathers

Consider Tiffany

THE "BROWN BESS"—WITH BAYONET
IN THE FRENCH AND INDIAN WAR: 1756

Consider was Humphrey's great-great-grandson. I have a copy of the *Journal* he kept while on patrol from August to November in 1756. A Sergeant, he marched as far as Saratoga, New York with his Company of forty-seven men before being taken with "Camp Distemper." He was sent home in a wagon.

Consider remained a staunch supporter of England during the Revolutionary War. Confined to his home in Connecticut for two years, he was finally released in 1780. He died in 1796, leaving a good estate, including a sixty-four volume library, extensive for that time.

He remained a member of the Church of England for his entire life and a committed Conservative.

The "Brown Bess"

The military rifle, familiarly known as the "Brown Bess," used in New England during the French and Indian and the Revolutionary Wars weighed from ten to sixteen pounds and was sometimes over five feet long. Still a smooth bore, not yet that accurate and loaded at the muzzle, its firing mechanism had improved to the flintlock system. As important, the hand held pike had evolved into the bayonet attached to the end of the musket. *Voila!* My male forefathers could fire their weapons at an enemy, then immediately charge with fixed bayonet, killing now possible both at a distance and at close range with the same weapon. Military tactics of the day centered on this double effort with musket fire followed by a bayonet attack, little different from the massed charge of the Phalanx in classical Greek warfare in the fourth century B.C.

Weapons of My American Forefathers
Solomon Lombard
"The Brown Bess" Continued
The Revolutionary War: 1775–1783

Solomon Lombard was the descendant of my first American ancestor, Thomas Lumbert, who settled on Cape Cod in 1637. Solomon was born in Truro, Massachusetts on lower Cape Cod in 1702. His house sat at the edge of a marsh overlooking Cape Cod Bay, the trail to the Boston Turnpike nearby, still a primitive wilderness land.

Solomon attended Harvard in the 1720s. In 1750, he helped settle Gorham, Maine, the town's first minister. Bitter, unsettling warfare with the French and Indians characterized the world in which he lived.

He participated fully in the Revolutionary War, writing, of "the severe and Crewel [sic] Treatment that America has met with from the Court of Great Britain." The British troops should receive "blood to drink in full measure." Expelled from his Church for his liberality, he served on the Committee of Correspondence, as a Judge, in the State Constitutional Convention, and the House of Representatives. His son fired Maine's first shot in the Revolutionary War.

At seventy-five he volunteered for combat duty to fight the British.

He died at seventy nine.

Another "Brown Bess"

The military weapon he and his peers carried was the same as that of Consider Tiffany: the "Brown Bess."

Weapons of My American Forefathers
Peter Zachary
THE KENTUCKY LONG RIFLE AND THE BOWIE KNIFE
THE REVOLUTIONARY WAR FOUGHT IN THE SOUTH

I believe Peter Zachary to be one of those settlers whose ancestors migrated to the American colonies from failed Scottish–Irish rebellions against an English King. He is first mentioned in my family history as petitioning for fishing rights on the Rappahannock River in Virginia in the1770s. Sometimes in his migration south he participated in the Revolutionary War, far more brutal than the one fought in the north. By the late 1700s he homesteaded land in southern Georgia, then his grandson moved to Alabama. The land, which the grandson homesteaded in Alabama by grant from President Monroe, had been taken from The Creek Indians in bitter battles at Burnt Corn, near the homestead, and at Fort Mims and Horseshoe Bend.

The Kentucky Rifle and Bowie Knife

The combination of slavery and constant, bitter wars against "The Five Civilized Tribes" of Indians caused the south and the southwest to be far more violent than New England or the Middle West. Neither Peter Zachary nor his descendants could have conquered without a weapon far more accurate and deadly than the musket: the Kentucky Long Rifle. The technique of rifling was brought to the United States by our European ancestors. American genius with technology made rifling practical. Rifling—cutting of twisted steel channels or grooves into the steel barrel—gives the bullet a twist as it is fired and leaves the muzzle. That spin makes it possible for the bullet to be precisely aimed at a target hundreds of yards away. This technological breakthrough was completed by a more refined method of flintlock firing. For the first time in history men could kill what they could see even though hundreds of yards distant! Though expensive, hand-made for each client, it was the preferred weapon of the frontier, helping win the battle of Kings Mountain in North Carolina in the Revolutionary War, helping kill and destroy "The Five Civilized Tribes."

The Bowie Knife was a vicious killing weapon, used at close quarters in murderous personal combat in such vendettas as Sand Creek.

Weapons of My American Forefathers

Stephen Lombard and James Lombard

Plains Rifle and Colt Revolver

The Texas Revolution against Mexico

The Taking of the South and the Southwest

Stephen Lombard and his brother, James, descendants of Solomon, left Maine for the south sometime in the 1830s: Stephen to lumber in Mississippi and his brother to Texas, where family legend places him as a ship captain. He was the beneficiary of the Texas rebellion against the Mexicans in 1836 for, by 1839, he owned land in Brazoria County, Texas.

After Stephen's death in Mississippi in 1848, crushed by a log, his brother returned to Mississippi, married Rebecca, Stephen's widow. Stephen's only child was my great grandmother.

The south and the southwest in which they lived grew more violent as Texas rebelled against the Mexicans, as the United States conquered Mexico in 1846, as anti-abolitionist forces attacked the anti-slavery movement with increasing virulence.

Plains Rifle and Colt Revolver

Two improvements in weaponry accompanied this continual threat of violence on the southwest frontier: a more accurate rifle with greater killing power than the Kentucky Long Rifle and a repeating weapon. The rifle evolved into the shorter, heavier Plains Rifle, of rugged construction, delivering a larger ball with greater shocking power at longer range, fired by percussion cap. The repeating revolver was far more successful than the repeating rifle at first. Samuel Colt with his six-shooter was not only one of the repeating pistol's inventors, as important, he became its pre-eminent salesman, one of America's earliest hucksters and advertising tycoons.

For the first time in history man could now kill in succession!

And did so with ever increasing intensity.

Weapons of My American Forefathers

Bela Bentley Tiffany and William Bigbee Green

THE .58 CALIBER MUZZLE LOADING RIFLE IN THE CIVIL WAR

KILLING AT A DISTANCE IMPROVED

I had forefathers on both sides in that war. On my father's side, Bela Bentley Tiffany, Humphery Tiffany's descendant, was almost killed at the battle of Cold Harbor in 1864, fighting for the Union as a volunteer.

On my mother's side, the Greens of Alabama, descendants of Peter Zachary, served throughout the War as enlisted men fighting for the Confederacy. William Bigbee Green, my great grandfather, and six of my great, great uncles rode home from the War. They watered their horses five miles from the family farm. One uncle told the others to go ahead: he wanted to sit a moment by himself.

He was never seen again.

The .58 Caliber Muzzle Loading Rifle

Types, sizes, kinds of weapons increased in almost geometric intensity in the Civil War, the nation never to be the same again. The basic weapon was the .58 caliber muzzle-loading rifle though, by the end of the war, other rifles such as the Enfield and the Henry repeater had become weapons of choice. The Sharps gave us the name "sharpshooter," a sniper's weapon. The repeating pistol continued to improve. Primitive machine guns were invented. Artillery fire rose to new pitches of awesome killing power.

Yet...just as in the Revolution, the bayonet charge after the first volley remained a favored tactic, especially with southern forces, resulting in barbaric numbers of casualties.

Trench warfare even became a staple in the summer and fall of 1864, precursor to World War One.

Weapons of My American Forefathers
Edward W. Wood and Benton Chamberlain Wood
The Airplane
World War One

The plane Dad flew.

My father enlisted in the Army Air Service in 1918. After six months' training, he was made instructor and sent to Carlstrom Field in Florida.

Before being discharged in 1919, he flew the early Airmail, under contract to the Air Service.

I grew up with all his paraphernalia from World War One around our house: uniform, cavalry boots, spurs, hats, even some old artillery shells he or his brother acquired in France.

Once Dad cracked up. All his upper front teeth were on pins after he slammed his face into the dashboard. I still have in my house the instrument panel from that crash and the tip of the propeller from that airplane

Dad was the romantic figure in my life, his service as a pilot symbolizing for me the Knight of the Holy Grail of the medieval era.

The airplane was improved and developed during World War One.

The speed of the plane Dad flew averaged around a hundred miles per hour. There were no parachutes. The pilot flew out—or crashed.

The airplane introduced a total new quality into the nature of warfare: the capacity to kill civilians at a distance, without sensing their pain or the physical destruction of their cities and landscapes.

Dad's career as a pilot was, perhaps, the last era of romanticism in war.

On the ground in Dad's war the machine gun and poison gas intensified the horror…and, still, men used the bayonet in desperate moments of attack.

Dad's brother, Benton Chamberlain Wood, served on that ground, a corporal in the Army, battles fought by Americans in the same terrain I would penetrate twenty-six years later.

Dad's airplane evolved into the Enola Gay carrying the atomic bomb to Hiroshima, the German Luftwaffe bombing London, British and American planes decimating German cities, Curtis Le May's B-29's fire-storming Tokyo.

Dad's uniform on display in the Air Force Museum,
Wright-Patterson Air Force Base, Ohio.

The men Dad flew with.
Dad's in the back row, third from left.

CHAPTER TWO

✯ ✯ ✯

The Voices of My Forefathers

The Enemies They Confronted

So many nights when I cannot sleep, as I lie rigid upon my bed, overwhelmed by the violence of the day, it is the voices of these forefathers I imagine that I hear. Sometimes soft and gentle, sometimes harsh and overbearing, these are those voices of that long line of citizen soldiers who helped make the nation into all it has become. We murmur to each other, our link the wars we have fought, the battles we have known.

The romance of war and weaponry: Dad as a pilot in World War One.

They tell me of their triumphs; they tell me of their shame.

We share truths only those who know the hard duty of war and battle understand, these, then, my comrades of the night.

They ask with me, from beyond the grave: where does the violence in which we live come from? Where this hate? Where these enemies who recycle generation after generation? Why do we make new enemies each year, invent more powerful weapons to kill these enemies? Kill them now from enormous distances, free of the victim's pain? Why were we as young men taught to kill, shamed for the rest of our lives? Why does violence return in weary cycle? Why so virulent in our day? TV programs? Video games? Rapes? Murders? Wars? Children killing children, worst of all obscenities? This war on terrorism, leading us into unknown ways?

I hear Dad's voice so many times at night just before I fall asleep.

I tried so hard to protect you, son. Your mama and I tried so hard to protect you. Teach you what the world was really about.

Those guns I gave you? Remember those guns I gave you? Those books? Remember those books?

What being a man was all about.

Courage in a world of enemies.

Remember those guns?

I taught you how to shoot as a boy just the way my Daddy taught me. And his Daddy taught him, all the way back to the beginning. Guns in America, they made us men.

Your mama and I never dreamed you'd get so bad wounded in the war. I remember when I cracked up back in '18, how bad it cut my face, knocked out my teeth.

It broke my heart; you hurt wors'n me.

And you changed so much when you came home from the war.

Angry, bitter at your mama and me after you were wounded.

Oh, son, my son, once you were so happy. Laughed so much as a child. Sunny. A sunny disposition. Made your mama and me so happy just to be with you.

But, now, all this stuff you preach: giving up the gun. Change the world? DO AWAY WITH GUNS AND VIOLENCE? MAKE THE WORLD A BETTER PLACE? Sometimes I think that head wound done made you crazy, son. Always'll be men with guns, ready to do you in.

That's what all your forefathers knew. Why, they carried weapons and used them in the making of the nation. Your line…our line…goes back all the way, even 'fore Humphrey Tiffany, back to the very beginnings of our land. Do you think we would have a country without those weapons?

You been hurt so bad already.

I hate to see you get hurt anymore.

You left a career. Worked as a gardener. Dirtied your hands. You weren't raised to work with your hands. Gentlemen don't work with their hands. Quit those good jobs, the firm you were Vice-President of, MIT. Quit to become a writer.

Like you're running all the time.

You hurt me so much.

Hurt your mama.

We love you, son.

Love you so much.

Oh, Eddie, you changed so much since you got shot. So much pain inside you. It hurts me and your mama so much.

You got to see it clear, son. You can't give up the gun. What would the nation be without the gun to protect itself from these terrorists, this murderous violence of your new century?

Without a gun you will be nothing in this world. Nothing.

It's all we men got to protect our women and children. The gun is all we got. There are so many bad people out there, son.

Without a weapon a man is naked to his enemies.

My father did his best to ensure I was prepared to live by his tradition. Not only those weapons he gave me as a boy, the memorabilia from his service as a pilot in World War One scattered around the house, but the books he gave me. I still have the fifteen-volume history of the United States, *The Real America,* I read as a boy. The story told is one of war, war as glory, the making of the nation, the proper duty of a man. Several families are followed over four hundred years of American history from the conquest of Mexico through World War One, the sons forming the new nation out of conquest and war.

And later when I turned twenty-one, Dad gave me the four-volume biography of the Confederate General Robert E. Lee by Douglas Southall Freeman, and Carl Sandburg's six-volume biography of Abraham Lincoln, qualities of American men forged in the Civil War.

I reached maturity out of a world of war, the myths that men must defeat enemies with the weapons of their day, their reason for manhood.

Dad intimated—he never really told me—of our forefathers before the Civil War, the men in our family who helped settle America, essential to America's Colonial and pioneer past.

Those New Englanders who helped found the nation: the Lombards and Tiffanys in Massachusetts and Connecticut—men who first fled England with John Winthrop; Humphrey Tiffany who helped decimate the Indian nations of the Northeast in King Philip's War of 1675; Consider Tiffany who fought in the French and Indian War of 1755–1763; and Solomon Lombard who helped form the nation in the Revolution.

Southerners who conquered the south and the southwest: Peter Zachary fleeing from an English King, his hard Scotch-Irish tradition rooted in centuries of Celtic violence. Then his children and grandchildren—the Watsons and the Greens—carrying conquest deeper south as they migrated to Alabama, helping to destroy the "Five Civilized Tribes," killing, killed in turn, driving the Cherokees to the Trail of Tears. Finally, there were the Lombard brothers who drifted southwest, searching for a fortune, one brother lumbering on the Mississippi River, another moving to Texas after its fierce rebellion against Mexico.

These, then, my American ancestors before the Civil War, their battles to take the land and make the nation, their guns, their enemies, the American tradition I inherit.

Of all these ancestors it is Peter Zachary who tells me the most about early America. I see him, fleeing those hard Scotch–Irish wars where his ancestors died in massive battles against the English, rebellion after bloody rebellion: O'Neill's in 1649, Sarfield's in 1689, Tone's in 1798, the Scottish battles of Bannockburn, Flodden, Pinkie, Killiecrankie Pass, Prestonpans, the final massacre of the Highlanders at Culloden in 1746, these Peter Zachary's harsh inheritance.

Map One

The Battles of My Forefathers Before the Civil War

The First Chorus

Being a Dialogue with the Voices of My Forefathers
Who Took the Land from the Native American
Before the Civil War

Peter Zachary
Humphrey Tiffany
Consider Tiffany
Solomon Lombard
The Lombard Brothers: Stephen and James

I see Peter Zachary, his Kentucky Rifle always ready in his arms, tall, thick-shouldered, a bull-like voice dominating all around him, hear him at four when I wake, fearful of the violence of my day.

You Daddy's right, boy.

Without a gun a man is nothin'.

He ain't free to defend himself, take what he wants.

Why, we took the land from the Creeks in the south then the Cherokee. Ol' Andy Jackson led us to take the land that was ours, filthy, dirty savages. There wouldn't a been no America if we hadn't a took that land. "Civilized Tribes!" They lived like animals. Didn't deserve their land. We took it from 'em with this!

He lifts his rifle triumphantly as Humphrey Tiffany steps to his side, the Puritan who helped destroy the Wampanoags in 1675. Dressed in a black long coat, knee britches, a flash of white linen at his throat, silver buckles on his well-polished shoes, he speaks.

Do you think we enjoyed it? Killin' all those Indians? Why, they had taken food from us... eaten at our table...their little ones, their women. But, suddenly, they descended on us in war, cruel, screaming savages. Raped our women. Killed our babies. Burnt our towns. Our men killed and wounded. It built a mighty rage in our hearts. The rage of Jehovah. We paid for our freedom in the death of men. Torturing devils. Watched them cut a man to pieces, fingers to the knuckles, hands to the wrist, toes, feet, legs...a heated rifle shoved up his rear. His screams...I will hear them until I die.

Consider Tiffany, white gloves protecting his hands, his coat fine broadcloth, the darkest green.

No different for us in 1756. Until we destroyed the Indians, then beat the French, took their land, we knew neither safety nor peace.

Solomon Lombard moves quickly to his side, his coat and tri-cornered hat the uniform of a soldier of the Revolutionary War. His voice is hard with his contempt for this man at his side, Consider Tiffany, a man who refused to support the Revolution against the British, a Tory. Tiffany, always the aristocrat, shrugs as Lombard speaks.

Guns weren't just about killin' Indians and takin' land in the Revolution but 'bout our freedom. We lived and died by the sweet words of Tom Paine, Ben Franklin, Tom Jefferson. They told us we could form a new Paradise upon the land, one never seen before. We used the gun and the bayonet, for sure, but for this new vision of how men could live, free of arbitrary and cruel rule.

Peter Zachary interrupts.

Without the guns there would a been no nation. I fought in North Carolina, kilt those God damn Lobsterbacks with my Rifle, brained them with the butt, shoved a bayonet in their guts. Took land after that, good sweet land in Georgia. My sons and grandsons went to Alabama, fit the Creeks for land. Kilt 'em at Horseshoe Bend with Andy Jackson after they kilt us at Fort Mims.

You don't know what it was like. I shot an Indian boy of four or five and didn't blink an eye. Before I kilt him, made him watch me while I raped his mama. Raped her in the rear. Tore her up good. Wouldn't treat her like no white woman.

My own grandsons, they got kilt in '36 a tryin' to move those murderous Cherokees out of the south so we could have their land.

That farm your great granddaddy owned, the one he come home to after gettin' beat in the War Between The States, why it come straight from violence, from us a takin' it from the Creeks. Jus' like ol' Massachusetts and Humphrey Tiffany here, we took the land from the Injun by force.

With this!

He shakes his rifle fiercely

James Lombard, his clothes those of the pioneer, deerskin leggings and all, Stephen Lombard, sweat pouring from his face, the whip of a driver dangling from his hand, both join Peter Zachary.

James

How do you think we took Texas from the Mexicans? They kilt white men at the Alamo and slaughtered 'em at Goliad. Texicans butchered them at San Jacinto. Dirty yellow-skinned bandits. Raped their women. Christ! Kilt them too. Mad with the pleasure of killing.

Stephen

You can't imagine what Mississippi was like, boy, in the 1840s. I followed my brother south...got caught up in Mississippi when I met your grandma. Worked in lumbering in Maine first but heard of all the money to be made in the south. Wandered down. Bought lumber land on the Mississippi River. Sold the lumber. Bought more land. Grew cotton in the Delta.

Mississippi!

Nobody understands it today!

A killin' and a fightin'...Mississippi jus' turned a State thirty years 'fore I come.

Slaves everywhere. So hard for you to imagine. A man had to have a gun...that's why I carried a pistol and this whip...had to use 'em sometimes.

James

Needed guns. A man had to be so hard to survive. All he had was hisself.

Stephen

And his family...

James

That's why I married 'Becca. After Stephen died, felt so sorry for her. A right cute thing, she had this baby, your great grandma, born after Stephen died, in 1849. No place to go...so scairt when I come to her. Her money 'bout run out. I had to marry her, look after her...my blood.

We both kilt men. I did one in Texas... if greasers are men. Then a white man on my ship. Drunk. Sassed me. Hit him with a marlin spike.

Stephen

I kilt two black men. They were drunk too. Came after me with Bowie Knives. Shot 'em dead. Think that's why that tree crushed my chest. A brother of one of those slaves let his pike slip so the end of a tree trunk slammed me in the chest.

James

Can't imagine it today. The sickness. Malaria. Typhoid. Small Pox. Yellow Fever. Cuts infected, swollen with pus. Scratch insect bites all the time. Chiggers! My God! Chigger bites drive you crazy.

Stephen

The blackness at night. No white folks for miles around.

James

The blackness of the niggers…worked 'em to death, white folks did, hated their color, an evil color.

Peter Zachary

Hell, a man couldn't be a man if he didn't carry a gun, at least a knife, ready to use 'em when attacked, ready, most of all, to defend his honor against his enemies.

Why, a man's honor was everything.

Everything he stood for wrapped up in his honor.

Went right back to our ancestors, those good men who died for Scotland and her freedom, fled to the Colonies to escape the British.

We knew of honor from our birth. From our Daddy's loins.

Look at those times, boy, the times in which we lived, look at the way it was for men in those early days. We didn't have no choice. So many enemies to fight, so many enemies to overcome.

We had to use our weapons, the weapons of our day.

We didn't have no choice in the evils we committed!

Look at our times, boy!

Understand the enemies we faced, enemies everywhere we had to defeat. Indians on the land, France, England, Spain, each one strugglin' to take our liberty.

Take a long, long moment to look at our past if you want to understand the present. Look at your ancestors, the enemies they faced.

AND BEAT! THE LOMBARDS CRY.

It ain't enough to jus' talk 'bout the weapons we carried. It warn't just the weapons we carried to take the south and the southwest that made us men. It was the enemies we defeated. That's what made us proud. It was our honor we had to defend. You got to understand our times: the things that happened to men in America to make us who we were.

The women we loved and tumbled. The women we married. Our children. The work we did.

Most of all the enemies we feared.

Our enemies helped determine who we were, 'specially as young men.

Our enemies differed for each generation.

Start by getting that clear if you ever want to understand our violence: men in the taking of America, 'specially in the south, were born and bred to defeat their enemy with the weapon of their day. It made us into who we were and into all we became.

Yes! It is these voices from my past, these voices I so often listen to just as I fall asleep, those voices of grandfathers, uncles, cousins who made this nation into all it has become. It is these voices that tell me they formed a nation before the Civil War by defeating enemies with the weapons of the day.

Defeating Native Americans

By 1865 the American Southeast and Northeast had been virtually cleared of Native American Nations. In continuing wars, from 1637 on, these Nations were were burnt out, enslaved, relocated, ruthlessly murdered, an estimated 130 to 140 nations.

Destroying the Environment

Another enemy was the land, its plants and animals, fish and birds. God had given mankind dominion over nature, the earth a resource to be used as men saw fit, largely for profit.

Defeating Foreign Enemies

Colonial Wars from 1637 to 1763 helped form our national character. Military skills so acquired helped defeat the British in the Revolution and the War of 1812. Spain is defeated in 1819, Florida ceded to the United States. Mexico is defeated in 1846–1848, after the Texas Rebellion of 1836. Texas, New Mexico, Arizona, California, parts of Colorado ceded to the United States.

Enslaving African Americans

In 1619 the first slave reaches American shores in Virginia. In the 1660s Colonial legislatures pass laws denying freedom to people of color, even if Christian.

Enslaving Mexican Americans

After defeating Mexico in the War of 1846, Mexican Americans were sometimes enslaved.

Defeating Enemies Within

Pilgrims fought Puritans. Settlers fought each other. In Shay's and the Whiskey Rebellions, settlers fought the new American government. Abolitionists and anti-abolitionists riot. Protestants burn Catholics out. The culmination is our great Civil War.

If I am ever to understand the violence of this land, the violence I so fear when children now kill children, when terrorists so threaten us with their madness, it is not just the weapons of my fathers I must search. It is the way we make enemies I must examine, enemies of all kinds, part of the American national character rooted in our darkest heart.

Chronicles of the Violent Events and Enemies My Forefathers Confronted

In the Era Before the Civil War

A Chronicle of a Few of the Violent Events and Enemies Confronted in the Era of Squire Humphrey Tiffany and Consider Tiffany: 1599–1787

1599	Spanish soldiers massacre Indians of Acoma Pueblo in New Mexico.
1619	First slave imported.
1634	Pilgrims fight Puritans for trapping rights in Maine.
1637	Puritans massacre and enslave Pequot Indians on the Mystic River, Connecticut.
1655	Catholics and Protestants wage war for possession of Maryland.
1656–1661	Puritans persecute and hang Quakers in Boston.
1664–1671	Maryland passes laws denying freedom to persons of color and other Colonies follow this precedent.
1675	King Philip's War.
1676	Bacon's Rebellion against Virginia's Colonial government.
1689–1697	King William's War.
1692	Persecution and murder of supposed witches in Salem, Massachusetts.
1702–1713	Queen Anne's War.
1712	Slave revolt in New York. African Americans defeated and brutally tortured.
1739–1742	War of Jenkin's Ear.
1739	Three slave insurrections in South Carolina.
1742	Bloody election in Philadelphia.
1744–1748	King George's War.
1745–1754	New Jersey tenant riots.
1754–1763	French and Indian War.
1763–1764	Paxton riots in Pennsylvania: Scotch-Irish settlers murder unarmed Indians.
1765	Stamp Act Riots.
1766	New York Agrarian Revolution.
1770	Boston Massacre of civilians by British soldiers.
1771	Battle of Alamance in North Carolina. Rebels seized and executed.
1774–1775	Terrorism against Loyalists explodes.
1775	Battles of Lexington and Concord.
1775–1783	Revolutionary War. 4,400 battle deaths.
1786–1787	At end of Revolutionary War civilians rebel against new government in Shay's Rebellion.

A Chronicle of a Few of the Violent Events and Enemies Confronted in the Era of Peter Zachary and the Lombard Brothers: 1788–1865

1794	Whiskey Rebellion against federal government over taxes on liquor.
1804–1806	Duel kills Alexander Hamilton. In Dickinson–Jackson duel, Andrew Jackson, future president, kills opponent who Jackson claims slandered his wife.
1811	Louisiana slave uprisings.
1812–1814	War of 1812 with England. 2,300 battle deaths.
1813–1814	Creek Nation destroyed and enslaved in Alabama.
1817–1818	First Seminole War in Florida.
1818–1819	War with Spain over possession of Florida. Spain cedes Florida to U.S.
1822	Vesey slave uprising in Charleston, South Carolina
1831	Nat Turner slave revolt in Virginia.
1833–On	Agitation for abolition of slavery increases each year. Anti-abolitionists riot. Violence over slavery begins to divide the nation. Race riots.
1834	Burning of Irish Catholic Ursuline Convent in Charlestown, Massachusetts.
1835	Murder of Elijah P. Lovejoy over his abolitionist newspaper and press.
1836	Battle of The Alamo. Texas rebels against Mexico.
1838	Riot against Mormons in Haun's Mill, Missouri.
1838	"Trail of Tears." Cherokees expelled from the south.
1840s–On	"Know Nothings" riot against and murder immigrants and Catholics.
1844	Riot against Mormons in Carthage, Illinois. Joseph Smith, founder of Mormons, and his brother murdered.
1846–1848	War with Mexico. 1,700 battle deaths. 11,500 from other causes.
1850	Fugitive Slave Bill passed. Violence as slaves flee vigilantes.
1851	Germans and Irish battle each other in Hoboken, New Jersey.
1854–1861	"Bleeding Kansas." Murderous fights between pro and anti-slavery forces.
1855–1856	Municipal riots. Vigilantes take power in San Francisco.
1856	Charles Sumner, senator, almost beaten to death on floor of Senate.
1857–1861	*Dred Scott* ruling by Supreme Court only exacerbates violence over slavery.
1857	Mormons massacre 120 American settlers in Mountain Meadows, Utah.
1859	John Brown attempts a slave uprising in Harper's Ferry. At his execution says: "I, John Brown, am now quite certain that the crimes of this guilty land will never be purged away but by Blood."
1861–1865	Violence of thirty years explodes in the Civil War, America's bloodiest war.140,500 battle deaths, 224,000 from other causes in the north, 74,500 battle deaths, 59,300 from other causes in the south.
1865	Abraham Lincoln assassinated.

The Second Chorus

Being a Dialogue with My Forefathers Who Fought Each Other in the Civil War

Bela Bentley Tiffany, a Volunteer in 1862,
36th Massachusetts Volunteers, and
William Bigbee Green, a Volunteer in 1861,
3rd Alabama Cavalry
and
Solomon Lombard of the Revolutionary War
With Unexpected and Undesired Interruptions
by Peter Zachary

Part of the reason we have become the strongest nation in the world lies in this strange inheritance, our ability to defeat enemies linked with incessant improvement of our weapons, used in murderous fashion, from the destruction of the Pequots in 1637 through our latest bombings of Iraq, Serbia, and Afghanistan.

Violence is not separate from our national character.

It is, sometimes, all we are.

I know of it by these scars pressed into my flesh from my wounding in World War Two, this right hand that never works properly, these emotional waves that threaten me.

I know it from the haunted years I wandered so alone, a marriage destroyed, children almost lost.

I know it down to my genes, essential to my inheritance.

Perhaps that is why I feel closest to those forefathers who fought each other in the Civil War: Bela Bentley Tiffany, who enlisted in the 36th Massachusetts Volunteers in April 1862 at the age of twenty-seven, and William Bigbee Green, who, with his brothers, enlisted in the 3rd Alabama Cavalry in September 1861, at the age of nineteen. Both Bela and William were wounded before the war ended: Bela Bentley at Cold Harbor in June 1864, and William Bigbee at Munfordville in September 1862 when the Confederate Army of the West invaded Kentucky.

A staff sergeant in the 36th, Bela Bentley fought at Knoxville and Vicksburg, then back to Virginia for the battles of the Wilderness, his life changed forever in the first seven minutes of the battle of Cold Harbor when he and 7,000 other Union soldiers were either killed or wounded.

His dedication to the Union and the abolition of slavery, willing to barter his life for those principles, his letters after that wound, his bitterness at the Copperheads in the election of 1864, and his fierce support for Abraham Lincoln, all tell me of a man I can respect.

And yet…I do not feel as close to him as I do to my great-grandfather, William Bigbee Green.

My lineage to this great-grandfather is through my mother, the most important figure in my life.

I have a photograph of his home, one of my mother, his granddaughter, then myself, on the great front porch of his home long after his death, finally a photograph of him, all taken in Peterman, Alabama, near Burnt Corn where the war against the Creeks began in July 1813.

I stood on that porch in recent years. Walked down the dusty road from his house past the schoolhouse my mother attended, the house where my mother was born in 1900.

That road, a mile long, those homes, that school house, all were once his property, deeded to the family in the early 1800s by President Monroe, in the same year Monroe promulgated the Monroe Doctrine, property taken from the Creek Nation.

I have the family Bible, listing my mother's birth in 1900 and his in 1842, only six years after two of his uncles were killed by Cherokees.

I own a few artifacts from his estate: two butcher knives that might have been models for the Bowie Knife, a powderhorn, and a chair for his wife or mother.

I feel this grandfather understands me in ways unknown to men of my own generation, unknown even to my father, who triumphed in his war of 1917–1918.

Map Two

The Battles of My Forefathers During the Civil War

After riding with the Alabama Cavalry for four long years, once wounded, William Bigbee Green came home to bitter defeat.

I was but a few months younger than he when I was blown up by artillery in 1944.

Eighty-two years younger than he, I feel we share the most common bond of so many men who have been in war and experienced violence: defeat.

William Bigbee Green, my Confederate great-grandfather, who rode with the Alabama Cavalry for four long years.

I hear his soft, southern voice as I fall asleep, the voice of my southern forefather who rode with the Alabama Cavalry for four long years in what he called The War Between The States, the toughest war the nation ever fought.

Eddie... Eddie, he calls, you're on the right track, boy, you just got a long, long way to go.

It ain't the way my great granddaddy, Peter Zachary, said it was, the stories he passed on 'bout takin' land from the Creeks! He turned war and killin' into a kind of glory, nothin' like what I saw at Fort Pillow in the War Between The States.

It was always that way for him, a killin' and a fightin', what a man should be.

But he never saw it's bad's I did.

Bela Tiffany, my great uncle who fought on the Union side in the Civil War, wounded at Cold Harbor, steps to William Bigbee's side and speaks for the first time. He has paced back and forth, listening intently. Six foot three, wide shoulders, a sergeant's yellow stripes on the sleeves of his blue uniform. He looks directly at William Green, then turns to me, speaks in a low, powerful voice, vital with authority.

No, son, your daddy ain't right 'bout living by the gun, none of these fire eaters are right.

I fought for the Union just as hard as they did in the Revolution and the Indian Wars, as Bill Bigbee here did fighting against me. Got shot for it at Cold Harbor. Just the way you got shot in your war, Bill Bigbee in his. Saw awful things that spring of '64. What we did, what the rebs… he nods toward Bill Bigbee… did, the way they charged us with those awful bayonets a glintin' in the late evening sun, the way a bayonet rips up the guts.

Saw us do the same.

We had to do it to win, why we fought, for the Union and to end slavery.

But that don't make the killing right.

Nothing makes the killing right.

All my life I brooded over what I done, the men I kilt.

Solomon Lombard, dressed in his Revolutionary War uniform, cocked hat, knee britches, joins them, stares thoughtfully at Bela Tiffany, interrupts.

No, that don't make it right.

It ain't that way all the time.

It just ain't that way.

Sometimes we got to use guns for a good cause.

I heard the war cry of Thomas Paine, devoured the Declaration of Independence. A foolish old man of seventy-three, even got so 'cited, tried to volunteer to fight in the Revolution. Said I was too old...and they was right! Always hated guns. Never could hit nothin' with the Brown Bess. Too damn heavy. Too big a load. Never hit where I aimed.

Believed with the Good Book: "Thou shall not kill!" So hard for me to change my mind, come to killin'. Hated it, though we had to do it for freedom! Found a way to serve without killin'. Served in the General Court. Helped write the Constitution of Massachusetts. A member of the Correspondence Committee. Confiscated land from Tories in Cumberland County, deeded it to good patriots.

Let my son, Calvin, do the killing for me. He fired Maine's first shot in the Revolution. Fired a musket at a British Man O' War anchored in Portland's Harbor.

I hated the killin'. The Good Book said it was wrong. But [he fiercely shakes his old head] *we had to do it. No other way to win our liberty from the British.*

William Green joins the chorus again, speaks directly to his great-grandfather, Peter Zachary, who still carries his Kentucky rifle.

No, grandpa, no, you ain't right 'bout this.

I worshiped you as a boy. Heard the family stories 'bout takin' land from the Creeks and Cherokees. Burnt Corn, Fort Mims, Horseshoe Bend. How we loved Andy Jackson. When the War Between The States come along, all I wanted to do was fight, me and my brothers. I kilt men, Grandpa, kilt more than you ever did.

Did awful things in the war and after it. Awful things. Brooded over 'em all my life.

Never forget the man I just shot looks at me with soft eyes as he dies. The way a woman cries after… he shakes his head and stops.

I know I felt awful as I got older 'bout the things I did as a young man. Awful. Usta lie awake at nights a frettin' 'bout the Yankees I shot, 'bout the black women I…oh, I did the same things as you, Grandpa, took colored women 'gainst their will when I rode with the Klan. Awful things to do. I…

Peter Zachary steels himself as he answers his great-grandson.

I never thought about it.

And you shouldn't either, boy. You did what you had to do to protect your family and your people. Those Indians, those blacks were nothin'. NOTHIN'! They didn't fit with us. Strangers. Enemies. Not worth a pot to piss in. Killin' 'em like squashin' a bug.

William Green interrupts and shouts.

NO! NO! Grandpa.

It just ain't that away. Jus' 'cause we had to kill as young men, that don't make it right.

I never thought I'd agree with an ol' Yankee, he repeats, his voice quivering with his pain, but like your Uncle Bela here, boy [he speaks to me], *I been studyin' the war we fought over states' rights and slavery for 'most a hundred years.*

And this uncle o' yours speaks the truth, young man. I know what killin' did to me. Made me hard. Unforgiving.

Bela Tiffany shakes his head in bemusement.

And I never thought I would agree with a reb. Maybe we both saw too much killing when we were young.

And worse things, William Green interjects.

And worse things, Bela agrees. Sometimes we have to kill for a just cause but that don't make the killing right.

My brother, Levi, ten years older 'n me, he never felt it was right to kill, even to save the Union and abolish slavery. Never believed that violence accomplished any goal. Believed with the Good Book, like you, Solomon, "Thou Shall Not Kill."

Wrote about being against war, the War of 1846 against Mexico. There's some papers in those boxes of family papers your Daddy left you. His writings are there.

He taught me that there is a different history of America hidden beyond our violence, behind our dedication to weapons and enemies, the killing that Peter Zachary so proudly salutes.

If you want to reach the truth, young man, discover a way beyond violence, you must find that other history. It is far harder to discern than your Chronicles of Violence and Enemies. Old men like Peter Zachary, like those who cry the Greatest Generation in your time, glorify war, and young men want to hear their stories, bloody battles, heroism in bloody combat.

But tell this other truth of men and women who sacrificed careers, families, even lives for a way beyond violence? Men and women who struggled to defend the rights of the poor, the weak, the people of color? These men and women are not heralded in our histories and you must search to discover them.

It will take much work.

And William Green speaks again.

By God, this ol' Yankee uncle of yours is just plain right. There's another truth about America you got to discover, besides the one you already found, the weapons and the killin', a truth hidden, a shining jewel in the muck. Dig for it, boy. Find it. In it you may discover what you search for, an ethic, a way of life beyond the killin'. I have searched for it for all my life since the War and the killin' I did in it, the hard things I did, the butchery at Fort Pillow, the nights when I rode with the Klan.

Sweet Jesus, he shakes his head, the evil young men do.

Peter Zachary interrupts.

Don't pay no nevermind to that Yankee!

Bela Bentley

Look for it, boy, the truth of America beyond our guns and killing. Out of an unknown and hidden past might yet come a new future.

William Green, glaring at his grandfather, joins Bela Bentley in final chorus.

Find that truth about America, discover it, show it to the world.

Bela Bentley Tiffany at twenty-seven when he enlisted as a volunteer for the 36th Massachusetts.

Bela's Letter

I cannot fight the enemy in the field I can fight them as
effectually at the polls at the coming election, As a Mass-
man I am proud of the honors shown Mass men everywhere
It is a passport to anything almost to hail from Mass
And as a Mass man I want to feel proud of the stunning
crushing majorities that shall come down on this miserable
patchwork of a peace party that is to detestable to think
about, I have already gained a No. of good friends here some
of the Ladies that will assist me all they can to get a furlough

Bela Bentley Tiffany's letter seeking the election of Abraham Lincoln in the Fall of 1864 (written after he was wounded that spring).

"I cannot fight the enemy in the field. I can fight them as effectually at the polls in the coming election. As a Mass [Massachusetts] man I am proud of the honors shown Mass men everywhere. It is a passport to anything almost to hail from Mass. And as a Mass man I want to feel proud of the stunning crushing majorities that shall come down on this miserable patchwork of a peace party that is detestable to think about."

CHAPTER THREE

✯ ✯ ✯

Discovering Another America

Voices of Those Who Lived
Beyond the Weapons of Our Fathers

While contention in the chorus continues
And cycles of violence rage

Yet, as balance to our violence, the peace movement in the United States began as a reaction to the horror of the war of 1812. A decade later the abolition movement started here and in England, demanding freedom for the slaves. Great pieces of social legislation have been passed throughout our history. We have changed so many cruel conditions. Freed ourselves from Colonial rule. Abolished slavery. Given women the vote. Improved working conditions. Passed health legislation. Created opportunities for ordinary citizens never given by any nation.

Here in this quiet study I search our history for evidence of these qualities in our character beyond our propensity for violence, beyond the weapons of our fathers and our capacity for "enemy-making." Hidden in works seldom read, tucked within the pages of dusty history books, I slowly discover the memory of events of peace, people of peace part of America's long history.

I write and read and search for the truth about the American National Character here in my basement kingdom, encircled by these memories of our past: artifacts and records of my family, books of our long history.

The bamboo fly rods of my childhood hang on the wall. Fishing reels my father and I carried on our canoe trips into the Boundary Waters of Minnesota when I was a boy dangle next to them. Deep-sea trolling rods, remnants of fishing trips to the Gulf Coast, stand in the corners.

Sailing gear from the Bulls Eye, the sloop I once owned, sit on the shelves: the brass light from the mast top, fog horn, even the anchor rode and sheet I spliced.

My great-grandmother's coffee grinder and coffee pot, the ones Stephen Lombard's daughter used for so many years in New Orleans, are on my desk; next to them Levi Tiffany's gold-headed cane; the 75mm artillery shell Dad or his brother brought home from World War One leans against the wall; Dad's flight book, showing his first solo, William Green's powder horn and butcher knives, stacks of old 8mm movie canisters, recording the faces and places of a once great family—Chicago, Mobile, the Gulf of Mexico—all these are scattered on the book shelves.

In the furnace room next to my study lie all the family papers, inherited from my father, going back to the 1630s, southern and northern branches. They tell of ceaseless migrations north and south, south and north, papers that changed the pattern of my life after my father died, even family Bibles inscribed with the birth and death of family members.

Inheriting that history of a great American family forced me to search for the meaning of my past, record its complexity. As an only child I had been given a precious gift, a moment's light on the darkness of the past, a darkness I had to penetrate. My effort to understand, then narrate, the true experience of one American family, has formed and shaped my life for over twenty years, draft after draft of manuscripts finished, each a little closer to the bone.

And books at the side of these manuscripts and family artifacts, books strewn over the floor, by the bed upstairs, in the living room, on the kitchen table, books splashed with soup and dribbles of ketchup, books almost bankrupting me: Kirkegaard, Nietzsche, Becker, Marx, Camus, Lifton, May, Slotkin, Girard, Fromm, Jung, Freud, Dostoevsky, Sartre, Rank, Brown, Mumford. Histories of the United States: establishment, revisionist, little known discussions of the peace and social justice movements, their people, theologies of non-violence. I read novels and poetry and anthropology and sociology and economics: Shakespeare, Aeschylus, Sophocles, Euripides, Conrad, Tolstoy, Cavafy, Owen, Sassoon, Hopkins, Dickinson. I read at night and in the morning and, when I worked with my hands to earn a living, read at lunch.

Twenty plus years since quitting my profession to probe my past for understanding, to unravel the meaning of the puzzle my ancestors left me: on the one hand, ceaseless violence, on the other, a search for a life of peace and beauty. Read and wrote. Wrote and read. Traveled to Europe. Found the site of my wounding in World War Two forty years later to the day. Lived in Chartres, France near the cathedral, wrote the first draft of my book, *On Being Wounded*. Returned to Europe, bummed through its great cities: Madrid and its Goyas, Florence and its cathedrals, Rome and its monuments, Greece and Turkey and their classical cities.

Continued writing after *Wounded*, always seeking an understanding of violence, set within the frame of my family's story in America. A novel, tracing how one act of violence, a rape, spins down over generations. A long poem showing America's destruction of its land. Another novel, centering on pacifism at the center of war. Kept up my Journals while always searching for the meanings of war and peace, violence and compassion.

This effort was no academic pursuit, not a thesis for a Ph.D., properly documented, footnoted and approved by learned committees, but a penetration to the marrow of my soul and into the darkness of my past: what was the truth of my inheritance of enemies and weapons?

I finally discovered among all those papers my father left me the words of Levi Tiffany, Bela's brother, also my great, great uncle, about war. In one of those boxes written on thin, blue paper in the most beautiful of handwritten script, I read of his opposition to the Mexican-American War of 1846.

Levi's Peace Testimony

"If any man say he love God, & hate
his brother, he is a liar, & the truth is
not in him." Bible. Now, can any man
love his brother, & at the same time
be doing him the greatest possible
injury, & by seeking to take his life?
I think not. Yet this is true with
regard to all engaged in the business
of war – viz mutual hostility to each
other. Perhaps some would endeavor to
apologise for war, in certain circumstances
But I think it would be very difficult
to justify the practice, in any circumstances & most of all at the present
time by the United States.
Oct 8th 1847
Levi D Tiffany

Levi Darwin Tiffany's Peace Testimony written in opposition to the Mexican-American War of 1846.

If any man say he love God and hate his brother, he is a liar and the truth is not in him (Bible). Now can any man love his brother, and at the same time, be doing him the greatest possible injury, by seeking to take his life? I think not. Yet this is true with regard to all those engaged in the business of war—viz. mutual hostility to each other. Perhaps some would endeavor to apologize for war in certain circumstances but I think it would be very difficult to justify the practice in any circumstances and most of all at the present time in the United States.

Oct. 8th, 1847
Levi D. Tiffany

Levi Darwin Tiffany

I keep his photograph on my desk in my study, a link with a past of peace and compassion of my own blood. I have his Journals for many years, written as both farmer and teacher in the hard and rocky Berkshire Mountains of Western Massachusetts. These show me a kind man, a just man, a stable man, a man who rose at four and bedded at eight, adored his wife and worshiped his children.

Look at his face, his kindly smile,
the crinkle around his eyes…read
a page from his Journals, his dedication
to a family and a place…a
hero of his times,
unpraised, unsung.

He never served in the Union Army in the Civil War. Born in 1824, exactly a hundred years before my birth, he was thirty-seven when it began. Though I believe him to have been an ardent abolitionist, I also think him a pacifist who would not take up arms as his brother, Bela, did. He stayed on the farm and worked the land while Bela volunteered and fought, wounded at Cold Harbor.

Discovering a man of my own family who testified for peace at times of war becomes one of the more remarkable events in my life.

Levi and his testimony led me on a new search for different voices from our American past, that hidden history of which William Green and Bela Tiffany so movingly murmured. I began Journals, recording events of peace in my own life and the history of America. Slowly over the years I created Chronicles of Peace to balance my Chronicles of War and Violence. Unable to discover other ancestors in my family who so opposed war, I found the biographies of men and women who left their careers, gave up their families, sometimes even lost their lives as they struggled for peace, sought justice for the poor, the neglected, the oppressed. Slowly, far more slowly, I read their words, recorded their acts of courage in my Journals, their willingness to march to a "different drummer," seeking a life beyond the weapons of our fathers, those I came to call my People of Peace.

My Roll Call of People of Peace in the Era of Levi Tiffany

Beyond Enemies, Beyond Weapons

Thomas Morton

Persecuted by the Puritans of Boston because he "danced" with Native Americans under the maypole every spring, insisting they were his equals.

Roger Williams

Expelled from Boston for heresy, founds a church in Providence, Rhode Island for religious dissenters. Writes *The Bloody Tenet of Persecution*.

Ann Hutchinson

Another charismatic religious leader, also expelled from Boston.

Mary Dyer, William Robinson, Marmaduke Stephenson, and **William Leddra**

All hung on Boston Common in 1659–1660 because of their testimony as Quakers.

John Everitt

Of Maryland, one of the nation's first conscientious objectors, who said in the 1660s that "he could not bear arms for conscience's sake."

Those Pennsylvania Quakers

Including **William Penn,** who conceived and implemented a Colony without a standing army, maintaining a loving and forgiving policy toward the Native American.

John Woolman

One of those early Quakers whose Journal in the eighteenth century is one of the most important works of American literature, he opposed slavery, ill-treatment of the Native American, use of tax dollars to support the military, and careless neglect of the poor, defining many of the major social issues that would haunt America over its history.

Anthony Benezet

His Journals carried forward Woolman's tradition.

David Low Dodge and **Noah Worcester**

Founded the first peace societies in the nation, laying the foundation for both the American and the worldwide peace movements.

William Ladd

After the violence of the War of 1812–1815, formed the first national peace organization, drawing together in one group both those totally opposed to war and those who would support it in self-defense.

The Grimké sisters, Angelina and **Sarah**

Southern aristocrats who rebelled against their slave-holding tradition, moved north, became public speakers for both abolition and women's rights, fiercely speaking out their truth.

Lucretia Mott

Quaker, unseated delegate to the 1840 Anti-Slavery Society Meeting in London (unseated because she was a woman), organizer of the first convention for women's rights in America—perhaps the world—in Seneca Falls, New York in 1848.

Elizabeth Cady Stanton

Another delegate, who was a co-sponsor with **Lucretia Mott** of the Seneca Falls Convention, who went on to become one of the crucial leaders of the struggle for woman's suffrage in nineteenth-century America.

Coacoochee

Seminole chief who defended his tribe against the invading Army of the United States in 1841 with a speech of deepest sorrow, set in poetry.

William Lloyd Garrison

Sparked the abolition movement from its beginnings in the early nineteenth century, founded the *Liberator* newspaper, the Abolition Society, wrote the *Declaration of Sentiments* for the New England Non Resistance Society.

Frederick Douglass

Escaped slave, self-taught writer, abolitionist, newspaper publisher, who was one of the few men to give support to women's rights before the Civil War.

Sojourner Truth

Also escaped slavery, became a leader of the Underground Railroad, bringing other slaves to freedom, wonderful speaker for abolition and women's rights.

James Russell Lowell

Author and poet, who mocked the Mexican-American War of 1846 with a long series of poems.

Ralph Waldo Emerson

An American original, resident of Concord, Massachusetts, creator of Transcendentalism, the belief that there are realms of thought lying beyond ordinary experience, intuition an important source of knowledge. He stressed unity with nature and also wrote on issues of violence and non-violence.

Lieutenant Ethan Allen Hitchcock

While in service vigorously protested American treatment of the American Indian and our invasion of Mexico in 1846.

Elijah Lovejoy

Abolitionist, killed while defending his printing press from destruction, in 1837 in Alton, Illinois.

Henry David Thoreau

Author, jailed for refusing to pay his taxes as a protest against the Mexican-American War of 1846, wrote *Civil Disobedience*, one of the most seminal essays ever written on non-violence.

Elihu Burritt

Founder of The League of Universal Brotherhood, a pacifist secular organization that sought the support of working men.

It is these People of Peace, these and so many more, who helped form a different America than the one of violence and enemies. It is these people who helped give us a vision of an America as a place of freedom, with equality for all. Their acts of peace and compassion, so often forgotten, lead us toward lives beyond the weapons of our fathers.

A Chronicle of a Few of the Acts of Peace-Making and Compassion Committed by People of Peace Between the Settling of the Land and the Civil War

1658	First hospital founded in private home in New York City.
1658	Richard Keene, a Maryland Quaker, fined for refusing training as a soldier.
1673	Rhode Island passes a law allowing for conscientious objection against war.
1688	Resolution of the Germantown, Pennsylvania Mennonites against slavery.
1709	First private home for mental illness established in Philadelphia.
1752	First general hospital in the nation founded in Philadelphia.
1775	Continental Congress passes a statute exempting conscientious objectors from military duty.
1775–1783	Members of Peace Churches refuse to serve in Revolutionary War as testimony to their conscientious objection against war.
1786	Virginia passes a Statute of Religious Freedom.
1791	Bill of Rights passed.
1815–On	Peace societies demand nations exist without war. Belief in non-violence essential part of the body politic.
1820	First State-supported libraries in New York and New Hampshire.
1830s–On	Anti-slavery agitation constantly increases as anti-slavery societies formed: marches, newspapers, political demands. Demands center on non-violent solutions at first.
1800s–1840s	A huge explosion of experiments in communal and utopian living occur following the Revolutionary War. Center on religious concerns, racial freedom, economic equity, and women's rights.
1820s	Free public schools become more common, with first free public high school established in Boston in 1821.
1830s–1850s	Edward Hicks expresses the hope of a generation in his series of paintings of the Peaceable Kingdom, the faith that the lion and the lamb will yet lie together.
1830s–On	American painters begin to paint the American landscape, see it for its beauty. Hudson River School founded.
1830s–On	An American literature is formed, unique to the new democracy.
1836	McGuffey's *Eclectic Readers* become common way for teaching students all over the nation for two to three generations.
1840s–On	Anti–slavery agitation increases.
1863	Emancipation Proclamation.
1861–1865	In spite of belief in abolition, some members of Peace Churches refuse to serve in the Civil War in both south and north.

Why, that's just the way it was!

Levi Tiffany, dressed in his old farmer's clothes, knee high work boots, a flannel hat, sweated brown at the rim, speaks.

When I was a young man in the 1840s, America was a nation of real believers, people everywhere speaking, crying, calling for a Paradise upon this new and fragrant land. Utopias everywhere. Frances Wright in Kentucky making a place where black folks were free, Garrison forming his Non Resistance Society, preaching non-violence. Robert Owen in New Harmony making a town where economic equality was sought. The Quaker artist, Edward Hicks, painting his Peaceable Kingdoms, showing us hope for Paradise.

YES!

America would create a Paradise upon this new land.

Oh, the land was so beautiful, so fresh. The perfume of wild flowers and grasses in the spring, why it swept over us and brought tears of joy to our eyes. The forests rolled toward the horizon, the streams and lakes so pure.

We knew, we believed that we could form a new world on this new land, a way of living never seen on earth before. In spite of slavery, in spite of Andy Jackson's Trail of Tears, we knew that America would give a new spirit to the earth.

I even breathed that spirit myself.

In the 1840s took the train, most exciting trip of my life, to Boston, the horse car out to Brook Farm in Roxbury where Hawthorne, Dana had come together to make their own utopia, and Margaret Fuller, Greeley, Emerson sometimes visited. They really hoped that men and women could create a new life of peace and justice, beyond the wars of Europe, a life beyond enemies and weapons. I stayed there a week, worked in the fields, talked all night 'bout our dreams of all America must become, a shining beacon to the rest of the world. Later, up in Concord, even talked with

Henry Thoreau 'bout that damnable war with Mexico 'fore he wrote his essay, Civil Disobedience.

It was Henry Thoreau's voice, the voices of all those peace-makers, that gave us vision, hope, a faith that in America this new life could be formed.

It is those voices that caused us to change the world.

Voices of Men and Women Who, as Peace-makers and Lovers of Early America, Helped Form a New Vision of the Land: Voices Levi Tiffany Heard

The more I looked the more I liked it. And when I had more seriously considered the bewty of the place, with all her faire indowments, I did not thinke that in all the knowne world it could be paralel'd for so many goodly grouves of trees, dainty fine round rising hillocks, delicate faire large plains, sweete cristall fountains and cleare running streams that twine in fine meanders through the meads,making so sweete a murmuring noise to heare as would lull the senses with delight....These people [Native Americans] are not, as some have thought, a dull or slender-witted people, but very ingenious and very subtile.

Thomas Morton, 1630s

There is a saying that we should do all...as will be done to ourselves....Here is liberty of conscience, which is right and reasonable; here likewise ought to be liberty of the body...to bring men hither...sell them against their will we stand against....

Resolution of the Germantown Mennonites, 1688

These Natives [Native Americans] have for [small] consideration sold their inheritance so favorably Scituated and in other places been driven back by superior force....And [English] people too often for the sake of gain open a Door for them to waste their Skins and furs in purchasing a Liquor which turns to the ruin of ym [them] and their Families....Wealth is attended with Power by which Bargains...are supported...and here Oppression...cloathes itself with the name of Justice....

Journals
John Woolman

...that the sinful disposition of men can only be subdued by love; that evil can only be exterminated from the earth by goodness; that it is not safe to rely upon the arm of the flesh, upon man whose breath is in his nostrils, to preserve us from harm; that there is great security in being gentle, harmless, long-suffering and abundant in mercy; that it is only the meek who shall inherit the earth for the violent, who resort to the sword, are destined to die by the sword.

William Lloyd Garrison

The land I was upon I loved: my body is made of its sands. The Great Spirit gave me legs to walk over it; hands to aid myself; eyes to see its ponds, rivers, forests, and game; then a head with which to think....The white man comes...they first steal our cattle and horses, cheat us, and take our lands...white men are as thick as the leaves...they come upon us thicker each year. They may shoot us, drive our women and children night and day...the red man's heart will always be free.

Coacoochee: Seminole Chief

I have been much with the Indians and look on them as part of the great human family, capable of being reasoned with and susceptible of passions and affection which, rightly touched, will secure moral results with almost mechanical certainty.

Lieutenant Colonel Ethan Allen Hitchcock

Under a government which imprisons any unjustly, the true place for a just man is also in prison. The proper place today, the only place which Massachusetts has provided for her freer and less desponding spirits, is in her prisons, to be put out and locked out of the State by her own act, as they have already put themselves out by their principles. It is there that the fugitive slave, and the Mexican prisoner on parole and the Indian come to plead the wrongs of his race should find them...

Henry David Thoreau

If you take your sword and dror it,
And go stick a fellow thru,
Gov'ment ain't to answer for it,
God 'll send the bill to you
Ez for war I call it murder,
I tell you straight and flat,
I don't want to go no furder
Than my Testament for that.

James Russell Lowell

That all laws which prevent women from occupying such a station in society as her conscience shall dictate, or which place her in a position inferior to that of man are contrary to the great precept of nature, and therefore of no force or authority

Seneca Falls Convention

That man over there says women have to be helped into carriages and lifted over ditches....Nobody ever helps me into carriages or over mud puddles....And aint't I a woman? Look at me! Look at my arm! I have ploughed and planted and gathered into barns, and no man could head me. And ain't I a woman? I have borne thirteen children and seen most of 'em sold off to slavery, and when I cried out with my mother's grief, none but Jesus heard me. And ain't I a woman?

Sojourner Truth

If you have a nation of men [sic] who have risen to that height of moral cultivation that they will not declare war or carry arms, for they have not so much madness left in their brains, you have a nation of lovers, benefactors, of true, great and able men

Ralph Waldo Emerson

These voices of peace-makers tell us of a completely different America than the one of Peter Zachary and his peers. Native Americans are seen as fully human, to be accepted into society, not murdered, raped, or relocated. African Americans and Mexican Americans are not to be enslaved; rather, abolition of slavery becomes society's major goal.

In the case of foreign enemies, great care is to be exercised in starting war. And, for some, there is not even this choice: they will not participate in killing, no matter the cause.

In some magical way these people saw no enemies.

They are the ones who saw the hidden motivations underlying our Wars of Imperialism and had the courage to say that truth, no matter the consequences to their personal lives. They are the ones who ask us to control the use of arms. They are the ones who, when enemies are real, sometimes help us modify our rage, limiting the damage unrestricted anger does to our national psyche.

Their common quality, besides their courage to struggle for their beliefs in a hard and bitter world, lies in their refusal to treat human beings as means to an end. Rather, the individual person is the end for which society is formed.

Our original democracy, at that moment in 1776 an enormous step forward in the history of human freedom, was for white men—propertied men!—only.

Actions of these People of Peace have played the essential role in broadening that circle.

In fact, over two hundred years of American history could be viewed as an awkward, lurching effort to bring an increasing number of groups into the democracy. Though, clearly, the effort is not finished—some would say, scarcely begun!—the original concept of democracy as a limited partnership has evolved into ever widening circles.

Actions of People of Peace have made this evolution possible.

Their minds resonate to the needs of others. They hurt when others are in pain. They long to change the causes of oppression.

These people, all, have a vision of human beings as able to construct a society of both freedom and responsibility where one is free to be oneself yet, simultaneously, has a duty to the greater community. Their view of human nature is hopeful: people can change, grow beyond themselves, solve old evils. It is in their compassion, their love of nature's beauty that I must discover my ethic for living in these violent times, beyond the weapons of our fathers.

* * *

Levi Tiffany, with tears in his eyes:

That's what so moves me, boy. You, my nephew three generations removed, are discovering the same words and actions that so shook me. Nothing dies! One act of compassion, of caring, of defiance ripples down through time in ways we will never understand. My words, those words I wrote in such utter despair, over the Mexican-American War, sat in old chests for over a hundred years, then you discovered them and they moved you, changed you! Doesn't that tell you something? That we must leave behind us some evidence, even the smallest scrap, of the best part of ourselves. Why, that scrap could change the world!

I was so much luckier than you. I heard all those wonderful people, even met some of them. In the 1830s and 1840s America was a place of such great hope and so many dreams: the lion and the lamb would yet lie together. We lived in peaceful, farming communities. Stable. Hard work every day but no TV, radios to distract us. Lived in the calming voice of nature. Friends we had known all our lives. People we could depend upon. Depended on us. Our families encircled us: mothers, fathers, brothers, aunts, uncles, cousins. We lived of and for them. Went to town meeting. School committee. Paid taxes. No violence. Never locked the door. Nobody ever got murdered or robbed.

Had good books to read. Took Harper's Weekly. *Went to Cambridge for plays. Why I remember seeing* Oedipus Rex *as one of the great nights of my life.*

Even with all that terrible violence in the Civil War, the killing of the Indians, the lynchings and the terrible slavery in the south, we still struggled to live peacefully, oppose the violence of the land.

And, now, you my nephew so many years ahead, have taken up that same goal.

It moves me mightily.

William Bigbee Green steps to Levi's side.

It moves me too for that's the way I finally learned at the end of my life.

That's the way we all try and live by, ain't it? Try and live without enemies? You done built a new life beyond enemies. Levi here did it early, smarter than me. Bela, [he nods at Bela Tiffany] *come to it too, didn't ya?*

Bela nods.

It took me a lifetime 'fore I got over the hatred coming out of the War. But I did it. Did it 'fore I died.

William Bigbee

Took me jus' as long, maybe even longer. Livin' in nature, tenderness, beauty...those all we can live by in times of great despair, all we have beyond the violence.

But the boy here, out of his war he come to it earlier.

That thing you wrote in your book about being wounded, you said it for all of us:

Being moved by ambulance from Battalion Aid Station into World War One bunkers, I smelled the cloying odor of blood, feeling it drop on me from the stretcher above. Here, at the front, before there were forms and bureaucrats to describe and fasten to me some categories of wound and pain, I touched men who, for an instant, cared for me with greater compassion than I had ever experienced at any time or place in my life—men who lit my cigarette and held it in their fingers while I puffed, men who returned to me again and again, seeing if my paralyzed hand would yet move, it my bowels were free, my urine not pink with blood, murmuring to me gently, even crying. I remember one boy my age who had ridden to the front with me, weeping as he cut off my blood-stained uniform at Battalion Aid Station. I remember the soldier I had watched being wounded visiting me in the field hospital, his arm still bloody in its sling, offering me a pack of cigarettes.

Gentleness and compassion, a softness so difficult for the American male to express—always there but held back, contained by some impenetrable shell, breaking now and warming me after I was wounded, as if love could only be given within the frame of violence and one must be expressed in conjunction with the other.

William Bigbee

That's the best thing about men you ever wrote, boy...we all got that tenderness and that love, us men. Oh, women don't think so but it's always there, the tenderness those soldiers felt for you, the tenderness you feel for your children and grandchildren, for the women you have loved, for your friends.

That tenderness lies at the root of your search for a way beyond violence.

In that tenderness and compassion life's meaning is discovered.

Bela Tiffany

And in beauty, the beauty of the America Thomas Morton described, the beauty of its earth and sky, forests, deserts, the beauty I found when I discovered photography after the war, opened my studio.

William Bigbee

The beauty I found on my farm, the light at dusk, the songs of sparrows, the pull of a horse when I plowed.

Bela and William together

What men with luck learn after they've been wounded in their war.

Like ol' Henry Thoreau: we march to a different drummer.

Bela

And if you can live this way, in nature, tenderness, and beauty, then maybe then, just maybe then, you got a chance to make some changes in this ol' world.

NOW WAIT A MINUTE!

Peter Zachary and his cohorts—Humphrey Tiffany, Consider Tiffany, the Lombard Brothers—have been standing at the side, impatient, their anger obvious.

Now, wait a minute, [Peter Zachary repeats].

You ain't even told half the story.

Your People of Peace done made all them fine pronouncements back in the 1840s, then look what happened! The fight over slavery bust over the land. Fugitive Slave Act. Riots. Raids. Murders. Killings. Murders in Kansas. Dred Scott decision. John Brown's raid at Harper's Ferry. Even your heroes, Garrison and Douglass, come out for violence to free the slaves.

THEN WAR!

BLOODY WAR!

The violence and the killin' always comes back.

Always enemies out there to be squashed!

WITH THIS!

Again he waves his Rifle triumphantly.

We always need our weapons to kill the enemy of our day.

ALWAYS!

ALWAYS!

ALWAYS!

Those People of Peace make no nevermind.

Look at the history the boy's done put together.

SURE YOU GOT THESE PEOPLE OF PEACE!

Sure they nice as apple pie.

But they don't make no nevermind in history.

The hate and the violence, the enemies and the weapons they always come back.

You got to be ready to kill.

And have the guns to do so.

His coterie—Humphrey, Consider, and the Lombards—nod their agreement and speak in chorus

The story being told of America jus' ain't the way it was. If we had followed those People of Peace there wouldn't have been no country at all. No thirteen original States. No Florida. No Colorado, New Mexico, Texas, Arizona, California.

The gun made us into the nation we have become.

Humphrey Tiffany

Why you still don't understand, do you?

I come from London in 1660, went to Swansea, part of Rehoboth. Had fled England, sold my land. Couldn't stand it when Charles II was 'bout to become King. The return of sloth, greed, lasciviousness. Cromwell was long dead. Our dreams crushed. I fled to this New World.

It terrified me. Cold in winter. Chill always in the bones. Rain. Fog. Snow. Bugs. All kinds of terrible bugs. Blackness of the forest. Redness of the Indian. Wild animals: bear, mountain lion, wolf, all strange monsters that could attack and kill.

Indians most frightening of all. Why, in '63 just when I come to Swansea, had to complain to the General Court in Plymouth about one of those thievin' savages.

Their women even went around with breasts hanging out, loins exposed.

Disgusting. Made my flesh crawl.

We had to be nice to 'em.

Had to.

Taught us a lot.

When Philip and his murderin' savages burst on us in '75, I was so lucky to escape. Had sent my wife away to have our baby but those savages 'most kilt my young 'uns. Did kill my neighbor, raped his women, blackness of the human soul, the Devil Incarnate, what they did to little children. But by the force of the Lord God Almighty, we trapped 'em in the Great Swamp. Caught Philip in Bristol, cut off his head, posted it in Boston after we drawn and quartered him.

You people today talk like we done wrong.

What else could we a done and made a nation?

Been like that Thomas Morton? Danced with Indians under the maypole? Rutted in their filthy beds?

That's why we left England, escaped Charles II and his lust.

In the name of Christ!

What else were we to do?

Bloody, filthy savages our enemies.

We kilt 'em.

Sold 'em into slavery.

WE, [he draws himself up proudly, waves his arm to include Zachary, Consider Tiffany, and the Lombards] *WE TOOK THE LAND YOU NOW CALL YOURS AND YOU BLAME US FOR THAT?*

Peter Zachary

> *Your daddy's right, boy. Without a gun a man, his nation, are nothin' in this world. Without better weapons every generation, defeat is sure. Your Chronicles of Violence and War tell us the truth of the nation, not those acts of peace!*

As I hear the angry voices of my forefathers, as I think on the violence of my day, I slide into a troubled sleep, a sleep of savage dimensions, a strange nightmare where I am imprisoned in a cell so small I can neither stand nor lie. Its black and damp-dripping walls slant toward me, their dark surfaces pocked with the barrels of all the guns, the weapons of my fathers: Arquebus and Snaphaunce, Brown Bess and Plains Rifle, Colt Repeating Pistols and the .58 Caliber Rifle, Lewis Machine Gun, Springfield Rifle, M–1 Garand. Their barrels swing toward me, follow me as I try and flee. They lower, aim at my soft genitals, dark holes ready to explode.

Do I hear the guttural voices of those ancestors, sharp with their contempt? Does one cry "*Fire!*"?

I cup my soft genitals with my hand.

I wake.

Soaked with sweat.

No wonder I long for a world beyond the weapons of our fathers, their enemies, their Chronicles of War and Violence.

A Chronicle of a Few of the Violent Events and Enemies Confronted in the Era of Levi and Bela Tiffany and William Green: 1866–1894

1866–1877 Reconstruction in the south, bloody retaliation against African Americans, formation of the Klu Klux Klan, repression of black rights as citizens by intimidation, beatings, rapes. Murders.

1866–On Lynchings become the American method for controlling African Americans, some 4,951 recorded between 1882 and 1927 though no one will ever know the number of deaths in dusty fields.

1866–1890 Wars against the Native American in the west explode in intensity. Veterans of the Civil War, generals such as Sherman and Sheridan, lead the American Army in destroying and rounding up Indian nations. Earlier in 1864 at Sand Creek in Colorado the militia, led by a Colonel Chivington, who said of Indian children: "Kill 'em, nits make lice!" had ruthlessly exterminated a tribe of unarmed Cheyennes. In 1890 the last Indian resistance was broken at Wounded Knee as American soldiers butchered unarmed Sioux.

1866–1877 Race riots continually explode, killing, raping African Americans as way to keep them in servitude.

1870s–On Rule of the "Gun" in the west. Rise of the myth of the gunfighter as essential to American history.

Violence against Chinese in Los Angeles. Anti-Catholic riot in New York. Police suppress marchers for labor rights in New York City. Great Railroad Strike practically stops all commerce in the United States. Army used as strikebreakers.

1878–1894 Labor violence all over the country: Chicago, Coeur d'Alene, Idaho. U.S. Army continues as strikebreakers.

1881 Assassination of President Garfield.

1884 Vigilante law in Cincinnati.

1885 Rock Springs, Wyoming, massacre of Chinese by Knights of Labor.

1886 Haymarket riot in Chicago with police killed by bombs, rioters by police. Anarchists hung by State.

1891 Anti-Italian riot in New Orleans.

A Chronicle of a Few of the Violent Events and Enemies Confronted in the Era of My Father: 1894–1945

1896	*Plessy vs. Ferguson* case decided by Supreme Court. Separate but equal doctrine becomes law of the land.
1898	Wilmington, Delaware race riot.
1898–1902	Spanish-American War. Atrocities committed against the Philippine Insurrectionists by American troops. Around 400 battle deaths and 2,000 deaths from other causes—but far more killed in the Insurrection.
1900s–On	US invades Central American, Caribbean nations on continuous basis.
1900	Boxer Rebellion in China. US sends troops to help defeat rebels.
1901	Assassination of William McKinley.
1905	Assassination of governor of Idaho.
1906	Race riot in Atlanta, Georgia.
1912–1914	President Wilson applies separate but equal doctrine to US Civil Service.
1913–1914	Ludlow, Colorado mine strike.
1916	Murder of unarmed WOBBLIES by vigilantes in Everett, Washington.
1916–1917	Punitive Expedition to Mexico.
1917–1918	US enters World War One. 53,500 American battle deaths, 63,000 deaths from other causes. Almost 10,000,000 killed or died from wounds worldwide.
1919	Race riot in Chicago, Illinois.
1919–1941	Lynchings of African Americans continue.
1919–1920	US Attorney General A. Mitchell Palmer leads federal government in arrest and expulsion of supposed "reds" from the US.
1921	Race riot in Tulsa, Oklahoma.
1922	Coal miners murder strikebreakers in Herrin, Illinois.
1931	Chicago tenants riot against eviction.
1932	Bonus Army of veterans from World War One expelled from their tent city by US Army led by Douglas MacArthur, Dwight Eisenhower, and George Patton.
1935	Violence in Arkansas against Southern Tenants Farmer's Union.
1935	Race riot in Harlem.
1937	Memorial Day massacre of unarmed strikers in Chicago.
1939	American Bund supports Hitler and Nazis. America First formed to bitterly oppose American entrance into World War Two.
1940	US institutes first peace time draft.
1941–1945	Pearl Harbor and World War Two. 292,000 American battle deaths. 115,000 deaths from other causes. Around 50,000,000 killed or died from wounds worldwide.
1942	US places Japanese citizens in concentration camps.
1945	War ends with atomic bombs in Hiroshima and Nagasaki.

In these years total war evolved into a horror we so easily accept today.
We no longer are appalled by massive killings from the air.
We no longer are even appalled by ethnic cleansings.
Our video games, TV shows, and movies are built around such events.
No wonder my dreams are nightmares of enemies and weapons.

Levi D. Tiffany speaks softly, his face troubled.

Enemies and weapons, these are the tragedies of America, so deeply part of who we have become that we no longer even realize their presence. All those wonderful People of Peace in the America of the 1830s and 1840s struggled so hard to live without enemies, without weapons.

Our times were so glorious in the making of America

So much, so much... his old eyes glisten with tears...of that hope has been lost in your era, as though people of your day no longer believe that, once, such dreams were real, not fantasies.

Struggle to understand why America has so many enemies today, uses these awful weapons to subdue them.

Bring this to the consciousness of the people.

Part Two—

Coping with My Inheritance of Enemies and Weapons

If I do not face these two qualities of my inheritance
I will never reach the place of peace I seek
Beyond the Weapons of Our Fathers…

CHAPTER FOUR

✯ ✯ ✯

The Persistent Quality of Enemies

With a Chorus Intermingled

On Monday, May 8, 2000, 135 years after the end of our Civil War, I randomly scanned the *New York Times*, immediately discovering headlines that revealed the persistent quality of the problem of "enemies."

Fighting Resumes Near the Capital of Sierra Leone
China Trying to Crack Down on Vocal Liberal Intellectuals
Europe's Migrant Fears Rend a Spanish Town

Shortly after that the *Times* began its long series on Race in America, pointing out the stubborn quality of racial bias, no matter surface and institutional improvements, the "other" still viewed as the dark shadow— the "enemy."

On May 17, 2001 the *Times* discussed road rage in an article on Los Angeles traffic, quoting a study saying that there may be as many as several billion "aggressive exchanges" between drivers in the United States each day, the driver in the other car seen as the "enemy."

On February 23, 2002, in *The Times Week in Review*, a headline read:

A Nation Defines Itself by Its Evil Enemies

On any day in any newspaper in the United States or on television news we are inundated with stories about enemies: The Balkans with their centuries-old ethnic hatreds; the Israeli-Palestinian conflict;

Catholic against Protestant in Ireland; Saddam Hussein in Iraq; ethnic murders gone bestial in Rwanda; racial rage in the United States—numbers of young black men in jail or on probation; serial murders, school killings, child abuse, rapes.

On any day on the Internet our Web sites spew out the literature and the images of rage, enemies everywhere: Blacks, Jews, Arabs, Gays.

At times events rising from our fear of enemies seem to peak: school shootings culminate in Columbine; terrorist acts escalate to Oklahoma City; a black man is dragged behind a car in Texas until his head is pulled off; a gay young man in Wyoming is beaten to death, crucified; a white supremacist in Illinois and Indiana goes on a killing spree.

Since the horrendous attack of September 11, 2001 all we have heard of is the "enemy," the "evil-doers" the bearded, dark bin Laden, his coterie, al Qaeda, now the Devil Incarnate, to be destroyed with our bombs, our "daisy cutters," that reek revenge.

In the years since the end of World War Two, it is as though our violence, our use of weapons to eradicate our enemies at a distance, with no risk to ourselves, has increased in geometric proportions.

A Chronicle of a Few of the Violent Events and Enemies Confronted in My Lifetime: 1945–1980

1946 Winston Churchill's Fulton, Missouri speech recognizes Cold War by his use of the metaphor, "Iron Curtain."

1947 CIA founded.

1946–1948 House Un-American Activities Committee runs roughshod over civil liberties.

1947 Truman initiates a Loyalty Program.

1948 Peace time draft initiated.

1949 Riot against singer, Paul Robeson, in Peekskill, NY.

1949 USSR explodes atomic bomb. Race for atomic dominance begins, ending forty years later with super-bombs, a stockpile of approximately 30,000 bombs, $4 trillion expended by the United States, thousands of explosions, earth, air, water polluted, residuals of poisonous, radioactive elements scattered over the earth.

1950–1954 McCarthyism dominant political force: civil liberties crushed in US.

1950–1953 Korean War. 34,000 battle deaths, 3,200 from other causes.

1952 First hydrogen bomb tested.

1953–1954 CIA helps overthrow left wing Guatemalan government, returns Shah of Iran to power.

1956 Hungarian revolt suppressed by USSR.

1958 US troops sent to Lebanon.

1961 Freedom Riders attacked in the south.

1962 Riot in Mississippi against James Meredith attending University of Mississippi.

1963–1968 Assassinations begin: Medgar Evers, Jack Kennedy, continuing with Malcom X, Martin Luther King Jr., Bobby Kennedy. Civil Rights struggle explodes.

1964 Summer. Three Civil Rights volunteers murdered in Mississippi.

1964–1975 Vietnam War escalates, a running sore on the body politic, a war never justified, a war that split the country apart, its wounds still scarring the nation. 47,400 battle deaths and11,000 from other causes.

1965 Civil disorder in the cities: Watts in Los Angeles.

1966 Summer. Forty three incidents of civil disorder.

1967 Summer. Civil disorder increase in intensity, peaks in Detroit.

1968 Assassination of Martin Luther King Jr. followed by three days of disorder. Vietnam atrocities peak with murder of civilians led by William Calley. Riot in Chicago at the Democratic Convention.

1969 Campus violence peaks. Movie, *Wild Bunch,* introduces a new level of violence into the media.

1980 The United States begins support of right wing forces in Central America.

A Chronicle of a Few of the Violent Events and Enemies Confronted in My Lifetime: 1981–2002

1983	Reagan sends Marines to Lebanon. 241 personnel killed in barracks bombing.
1984–On	Star Wars proposed as a national policy. AIDS becomes an epidemic. Movie and TV violence escalates. Murder and violence with guns also reach major levels. Militias begin to form.
1989	US invades Panama.
1989	RAP music increasingly expresses acts of violence.
1990–On	Child abuse increasingly reported. Repressed memory syndrome leads to arrest and prosecution of supposed sex offenders. Gang rapes in inner cities and suburbs reported in increasing numbers.
1990–On	Mass murders become commonplace: shootings in Los Angeles, on a train on Long Island, children killing children, media exploits O.J. Simpson trial, murder of Jon Benet. Black man dragged to death; gay man crucified. Shootouts in Idaho.
1990–On	Militias organize, protect themselves from enemies, survivalists.
1991	Gulf War. 148 battle deaths, 151 from other causes.
1991–1992	Beating of Rodney King in Los Angeles. After policemen found not guilty, civil disorder in Los Angeles.
1993	World Trade Center bombed in New York. Branch Davidian Compound in Waco, Texas destroyed by fire. Fire fight in Somalia kills 18 Americans and approximately 1,000 Somalians.
1995	April 19 terrorist bombing of Federal Office Building in Oklahoma City.
1995–On	Capital punishment intensifies in number of executions.
1999	US bombs Serbia. No ground casualties.
1999–2000	Another school killing in Littleton, Colorado. Two boys kill twelve of their peers, a teacher, and wound twenty-three others before committing suicide.
2001	World Trade Center destroyed, Pentagon attacked, over 3,000 deaths.
2002	War against terrorism begins with bombing of Afghanistan. Iraq, Iran, North Korea become "the axis of evil." War against Iraq possible. Israeli-Palestine conflict escalates.

And these are but a few of the events of violence at the turn of the millennium. Mass killings, continued capital punishment, stiffer sentencing of criminals, "three strikes and you're out," increasing crudeness of public life, lack of civility, money and consumerism apparently the major social goals, pregnancies outside of marriage matter of fact, welfare shifted to work, even for mothers with small children...and children kill children.

Throughout the world violence continues: Rwanda reeks with machete murders, Buddhists burn out Christians in India, ethnic groups in Yugoslavia destroy each other, Pakistan and India fight, gain atomic capacity. War between Palestine and Israel escalates into barbaric killings and destruction on both sides.

Why do efforts of our People of Peace so often fail? Why do these endless cycles of violence return each generation: we weep with our pain? What is there about human nature that endlessly repeats cycles of peace and war, violence and compassion? I have struggled for my entire life to answer these questions, understand the relationships among violence, enemies, killing, death, and the murderous and ever-improving weapons of technology. What happens when we become the "enemy," attacked by others simply for who we are, our innocents destroyed in shocking pain? Finally, over generations of war and violence, the failures of our efforts at compassion and at peace, the vaguest pattern slowly emerges, the darkest of shadows in the collective patterns of our human heart.

Qualities of the Enemy Deep Within the Human Heart

The Image of the Enemy

There can be no war, no violence without the existence of an Enemy...that person, thing, artifact we believe is "evil," out to get us or who possesses "things" we desire...persons so inhuman, obscene, so powerful a threat to our existence that we can only be saved by their total destruction: the Hun, the Jap, the Commie, the Indian, the Nigger, the Jew, the Terrorist, the Devil Incarnate, who must be ground to nothingness.

The Fact of the Scapegoat

The enemy is the scapegoat for our problems, our failures, our nightmares. When the enemy is destroyed, culture will be purged from evil. Justice, again, will prevail.

The Use of a Technology

The enemy is destroyed by the technology of the time from pike and bayonet to hydrogen bomb and "daisy cutter." Weapons technology improves each year in capacity to kill, more accessible to the "common man." Great amounts of economic resources are poured into the development and production of these technologies.

The Fact of the Hero

That act of destroying the enemy, purging the scapegoat through the use of existing weapons technology, is performed by "The Hero." He or she, who fights for the common good, kills the enemy, crushes the enemy into nothingness. Without enemies we cannot have heroes. (Are they, perhaps, made of the same stuff?)

The Fact of the Prized Possession

Our hero takes the prize from the enemy and returns it to us, the most precious of all human possessions. In war the enemies' city is captured. In peace the enemies' most valued property is taken.

The Warmth of Comradeship

We come together in union as we follow the hero in defeating the enemy. *Esprit de corps* is often formed from acts of greatest obscenity, torture and mutilation of "The Enemy" as the scapegoat who has defiled the Holy Grail.

These attributes of "enemy-making" recycle endlessly, confirm us in the righteousness of our violence, unconscious rites followed to destroy our enemies by the technology of the moment.

Not just in acts of war.

But in acts of repression in our ghettos where the poor, often people of color, are seen as less than human as we deny their children education, the food of life.

In our professional sports where, in football, sports now our national religion, teams are formed to defeat the opponent, young men injured, sometimes for life.

In our offices and our factories where career is more and more important to our being: we watch men and women snarl at each other like the animals they sometimes are.

In our families and our relationships where divorce lawyers tell horror stories of father against mother and mother against father, children so often lost in the struggle.

In our beds where now we demand orgiastic pleasure, often at the expense of our partner.

In our politics where, in Washington, one party seizes Congress, turns upon the other as the source of all evil.

In our language where four letter words automatically turn the other into the enemy.

Even in our imagination where, sometimes, the enemies we attack are not even real.

The terrible competition of our economic system exacerbates these conditions.

Our competitors become our enemies. We form teams—at work, in service organizations, in leisure time activities—to compete with other teams, sometimes, to grind them into dirt.

A larger part of our human stuff in America is devoted to "enemy-making" than we dare admit. The image of the enemy may help a person determine who she or he is. We meet the enemy at the edge of our personalities. They define who we are.

The enemy plays an important role in our search for freedom and justice. The enemy, the oppressor, the dark force in our mind, must be overthrown if the attributes of democracy are to be reached. Living always in our deepest memories are the wars we have won against an enemy, wars that brought us our freedom: the Revolution, the Civil War, Indian Wars, Wars of Imperialism, World War One, World War Two, the Cold War.

The image of the enemy runs amok in our video games where teenage boys spray that image with electronic bullets.

The image of the enemy forms the roots of our movies and our TV; the dark criminal and evil doer is overwhelmed by the "good" violence of the "hero," a kind of violence, of course, that has no real consequences.

The dark shadow of the enemy illuminates our nights in our TV mini-series, colors our days in newspaper stories, trembles in our dreams as we search for sleep.

We demonized Milosevic, now bin Laden and Saddam Hussein. In our minds they assume super-human qualities, symbolizing all human evil: get rid of them, smash them into nothingness and the world must become a better place, free of all suffering and fear.

Individual men have replaced the Soviet Union as the "enemy" we must destroy.

Even those boys at Columbine High School in Littleton, Colorado acted from their fear of and rage against their perceived enemy: the jocks who so put them down.

The process of turning the other into the "enemy" remains the same, whether the enemy is a nation or a person wearing the enemy's mask.

Supreme Court Justice Oliver Wendell Holmes, wounded three times in the Civil War as he fought for the Union and to free the slaves, posed the paradox of the enemy at the center of this book, when he wrote:

> *Between two groups [of men] that want to make inconsistent kinds of worlds I see no remedy but force...society has rested on the death of men.*

Holmes even carried his certainty about the need to defeat an enemy into the Supreme Court. In 1921 he turned the concept, "No duty to retreat," into the law of the United States. Common Law in England had stated that a person threatened by an assailant had "a duty to retreat," using every practicable means to escape the enemy who threatened his life. English law wanted to reserve the right of killing to the State, not allow it to the individual; only under the most extreme conditions was a threatened person given the right to defend himself, required, instead, to use every possible means to avoid conflict, including flight. After the Civil War, rulings of State Supreme Courts in Ohio, Indiana, Wisconsin, Minnesota, and Texas slowly rejected this concept. The "true man" theory took its place. The man who was a man did not run when threatened by an enemy. He stood his place and fought it out, killing his enemy himself, without punishment by the State for his act of self-defense. The case of *Brown v. United States*, 1921, made this concept the highest law of the United States, adopted by the Supreme Court by a vote of 7 to 2. If threatened a man could kill his opponent and be found innocent of the taking of life.

This law still rules our national ethos.

It is Clint Eastwood's "*Make My Day*."

In military practice I inherited the doctrine of unconditional surrender, total annihilation of the enemy in war, the rule that ended the Civil War, Bela Tiffany's and William Green's war, and my war, World War Two. That doctrine rose from the Indian wars; the enemy must be destroyed, eradicated, removed from the land, stamped into supine submission. Unconditional surrender when linked to "no duty to retreat" become the way men must deal with each other if they are to maintain their manhood, the "honour" of which Peter Zachary spoke.

My children, my grandchildren are taught these same beliefs.

They lie at the heart of our television programs and movies, the faith that violence breeds redemption.

At the core of American history, particularly since the Civil War, exists this tradition of violence, our fascination with "The Enemy" we must conquer. Without the enemy, that person, thing, resource of land or water or air, animal, bird, fish we must eradicate or control, there can be no violence, no ever more virulent technologies of death.

Seeing enemies everywhere, we came to these fresh and beautiful lands and waters and imposed a pattern of destruction upon the earth and its inhabitants: we killed the Native American, imported slaves, raped the land, eradicated the animals and birds, polluted the air and water, imposed these great metropolitan areas on the surface of the earth.

Dominance of our enemy lies in a dark and hidden heart of American life, causes of the violence we so experience each day.

Yet what real evidence do I have for what I believe to be a reality of the American National Character?

My southern inheritance, those men of slavery and the American wars.

The war in which I fought and was wounded.

The world we have created since 1945.

The testimony of so many people today that this moment in American history is one of fear, rage, insecurity.

The statistical evidence we have gathered about crime in the United States compared over time and to other nations. In terms of criminal acts against property, the United States is relatively similar to nations with like backgrounds, such as the countries of the United Kingdom, Germany, France, Canada, Japan, and Italy.

In terms of homicide, however, that act of permanently eradicating the personal "enemy," America still lies far ahead of other contemporary nations and even far outstrips its distant past.

The statistical fact is that homicide in the United States was far lower at the end of the nineteenth century, though it has steadily declined since 1994.

The statistical fact is that the homicide rate in the United States far exceeds that of comparable nations, for example those belonging to the G-7.

The statistical fact is that, in many ways, the homicide rate in the United States resembles that of Third World Nations.

The traditions of the southwestern frontier—my inheritance from Peter Zachary and the Lombard Brothers, the fact of the Civil War, "society rests on the death of men," combined with "no duty to retreat" form a central part of our National Character. The making of an "enemy," linked to the weapons needed to annihilate them, are essential qualities of our American heritage.

In the 1890s, thirty years after Bela Bentley Tiffany was wounded at the battle of Cold Harbor in June, 1864, the Tiffany family gathered for a reunion at the family farmhouse in Blandford, Massachusetts. Lydia Antoinette Tiffany Wood, my great-grandmother, came from Mississippi, her first trip home since the Civil War. Her brother, Levi, lived on the farm. Bela came from Indiana, Pennsylvania where he had his photographic studio.

I feel so close to all three: Lydia who migrated south to Mississippi, her letters reflecting the pioneer world where a rifle was essential to the protection of life; Bela Bentley because he too was wounded, knew that terror of combat, fought for a cause, the preservation of the Union, the abolition of slavery; Levi because he was a gentle man, a teacher, a farmer, and committed pacifist long before the word was coined.

So often I hear their soft voices as I struggle with my inheritance of enemies and weapons, my pain from the violence in which we live.

Levi, Bela, Lydia murmur among themselves. In the background I sense that William Bigbee Green hovers in the kitchen, uncomfortable in this Yankee home.

Levi, whittling a stick

You still feel the same, Bela? After all these years you still feel the same about the Civil War?

Bela

Of course I do, Levi. Course I do. I hated it, the killin', but it had to be done to preserve the Union and abolish slavery.

Levi

The cost? It wasn't too great?

Bela with great pain on his face

Oh, Levi! I think on it every day, the men I kilt. We had to do it and, if I were called again, there would be no choice. What else is a man to do when evil seizes the land? Justice Holmes was right. Remember when I went to Cambridge to hear him speak? "The death of men." But, oh, Levi, why, why must it be so? I hated the killing. You know that! Hated it! But to save the Union! To free the slaves! It had to be done.

Levi

But look what's happened. Black folks in the south are the same as slaves again. What's the name of that case in the Supreme Court? Plessy vs. Ferguson? Separate but equal facilities for black folks? You know that's a joke, Bela. Five hundred thousand killed for that. The Good Lord only knows how many wounded. For slavery to return in another form? Couldn't there have been another way 'sides the gun?

Bela

You're such a sweet man, Levi. You still don't realize what those southerners were really like. Secessionist devils! Deserved all we gave 'em. All! Why, I saw... [he pauses] *...what they did to black folks, the pain they made, we had to stop 'em. Now, Levi...* [he leans forward and puts his hand affectionately on his brother's knee]*...remember how much I believed like you at first? Hated violence as much as you did? Didn't believe in killin'. Believed the slaves had to be freed without war. But the south wouldn't let that happen. They woulda broke up the Union 'fore they freed the slaves. There are some men* [here his voice hardens, his fingers drum on Levi's knee]*...there are some men, Levi, and I saw them in the war, some men who are evil, just plain evil. If they are not stopped, they will destroy innocents. Our job in the Civil War was to stop them. We did. We kilt them, Levi. Kilt them. It was our duty under God.*

He pauses. His face strains, white beneath his beard.

I hated it. Hated the killin'. But when, in a man's life, there comes a time of evil, that evil requires a death, though the act of killing sickens.

Levi inserts quietly.

... and you were almost killed yourself.

Lydia speaks for the first time, her voice soft and gentle.

How we worried about you, Bela, all those years at war. That wound. We loved you so much.

Bela

The price we paid. The price any nation pays for its freedom. For justice. What our wars are all about. Freedom and justice. They both require the gun. There is sometimes an evil in the world, Levi, real enemies: only killing stops them.

Levi

What of God? Of Christ? The Sermon on the Mount?

Bela

Oh, Levi!

His old face softens.

I have prayed for forgiveness for all I did in the war, prayed for the men I killed, hoping they are in heaven.

I hated the killing.

Yet.

His face hardens again.

It had to be done.

I cannot answer for God anymore. I can only speak for man. Without the gun we used the slaves would never have been freed, the Union never saved.

I used the gun.

I killed with it.

It almost destroyed me, the killin'.

He pauses, turns to me.

But you did the same thing, boy. Your war, too, it was against evil, warn't it? The Nazis? No different from those slave owners. Treat people worse 'n dirt. Evil. Human evil. Sometimes it got to be stopped.

That's why men must use guns.

You ain't sorry are you, boy? That you did it? Volunteered. Went into the line. Had your head blown open, back and butt shredded. You ain't sorry?

A fierceness in his old eyes.

And in me a sudden exultation.

For I was not sorry.

At the core, the heart, lay this truth: some wars had to be fought, in spite of the killing, in spite of the cost. These two, the Civil War and World War Two, simply had to be fought.

And I was glad I had been there.

In spite of the wound, the years it had cost me, I did the duty, what a man was supposed to do.

And glad I could reach out and shake the hand of my great, great uncle, Bela Bentley Tiffany, over a hundred years of time.

Is this not the deepest link with the past a man can have, I wonder, that I was there, a man, fulfilling the lineage left to me?

But William Bigbee Green calls from the distant shadows of the barn, emerging from the kitchen.

Are you two fire eaters sure you're right? I believed as much in the cause of the Confederacy as you did in your faiths. But now I understand that I was wrong. The south was wrong. Slavery was an abomination. States' rights, the belief of narrow men who had no sense of the nation, only of their rights. Yet I fought for these beliefs as hard as you did for yours. I kilt for them. Sometimes men are just plum' wrong when they name another enemy and seek to kill. My cause was so bad! Christ! It pains. I kilt men for a cause proved worthless.

Lydia

And, Bela, what Levi said. Black folks aren't any better off since Reconstruction ended. Why down in Mississippi where I now live, it's…it's just awful, way whites treat blacks. Worse 'n slavery. At least as slaves they were worth money to whites. Now they worth nothin'! Nothin'!

Levi

And what of the costs of war, Bela? Not just men killed. But wounded. Wounded like you and the boy? Pain that spins down through time. What about the corruption and greed that rises from war, turns men into beasts?

Lydia quickly

Women corrupted too.

Bela stubbornly

The war was right. Freedom and justice. For any nation they are worth the death of men.

Levi and Lydia together

We don't know, Bela.

We simply don't know.

To know so readily when it is right to use violence?

To kill?

To know so easily what evil is?

To know when enemies are real? False?

You assume that humans possess the power to know all things.

We never had it.

Sometimes we turn our fears into enemies, fantasies, nightmares of the mind.

Sometimes our self-interest makes enemies of those who stand in our way.

Oh, Bela, we must be so careful. Once we make an enemy, there is seldom ever peace. The darkest shadows in our hearts overwhelm our charity and our love.

Bela

But brother, sister, you must know that from your life sometimes there are enemies out there, enemies men must defeat, enemies of the night. Even, he sighs wearily, Pete Zachary is right here.

AND YEAH!

I hear the brazen voices of that chorus, Peter Zachary, Humphrey and Consider Tiffany, and the Lombard Brothers.

Maybe you understand us now that you lost your own from a surprise attack, your innocents kilt. All your yellin' and bitchin' about our wars, the way we kilt those you call people of color. What we did was just the same as you when you bombed Afghanistan. King Philip's War, the French and Indian War, takin' land from the Creeks, the Mexicans, all that was jus' a strikin' back after we had been ambushed.

Humphrey Tiffany

The Wampanoags, why they come down on us, no different from those planes a tearin' into your buildings. We had to be as hard as we could. Hard. Hard. Hard.

Consider Tiffany

What the Indians did to our settlers, why there was no revenge that could fit the crime.

Peter Zachary

Five hundred settlers kilt at Fort Mims. By God, we had the right to kill at Horseshoe Bend.

The Lombards

The Alamo. Goliad. Americans butchered.

In chorus, all together

Dead or alive. We demanded justice for our innocents kilt, just the way you do, justice for the crimes committed against us.

And Bela murmurs softly.

Sometimes a man just got to kill his enemy for self-defense and for justice.

And ...he shakes his head...I hate it.

Peter Zachary gloats.

You be right. You be right. This World Trade Center attack, what they call "unique," think never happened afore, why it's common to our past. We get 'ttacked by strangers, we strike right back.

WE WIN!

WITH THESE!

He, Humphrey and Consider Tiffany, the Lombards growl and brandish their weapons again.

✯ ✯ ✯

The paradox of our time: why has American democracy over generations had so many enemies, many so clearly of the imagination? Are enemies thus an essential quality of our democracy, as important as the vote? To think about it in another way, why did our Colonial forefathers in Maryland declare through legislation people of color as slaves? What was there about the American experience on the frontier that led settlers to impose slavery on African Americans through acts of the body politic? Was it the terror of nature, the loneliness of the landscape, the very threat of each moment of the day? On the frontier there must always be someone to blame for our fears. The blackness of the "negro" personified the scapegoat. Does the pioneer experience inevitably lead to the formation of enemies as scapegoats, the need to destroy them, to have power over them?

The nature of the pioneer experience may also lie at the root of another paradox of our national character. Other nations elected to abolish slavery through legislation. England led the way in 1834 with abolition of slavery through acts of government. Even South American countries ended slavery through legislation. Yet the United States could not halt slavery through acts of government although an act of government, the Constitution, accepted slavery in the just formed United States in 1787. It took a brutal Civil War, fought in blood, to free the slaves. That war, forever, changed the quality of American domestic life, fixing attributes of enemy making into the American National Character: Unconditional Surrender and "No Duty To Retreat." Further, that war scarcely achieved the goal of freedom for the slaves. By 1877, twelve years after the south was defeated, African-Americans were forced back into a new form of slavery, separate but equal.

On the one hand our democracy demands that we accept the "other," the stranger and the migrant, as part of our body politic, "bring me your poor," emblazoned on the Statue of Liberty. On the other hand, that same stranger, that migrant, particularly when a person of color, is the enemy, who threatens our very existence, the dark enemy we fear in our unconscious, terrifying our days and nights, the enemy our suburbs are formed about. Whites flee their inner city image of "them," building gated communities, locked by income from people of color, by high walls and armed guards.

Perhaps the tension between such conflicting demands—accept the colored stranger? destroy him?—is too much for a nation in which there are so many immigrants, particularly since the Civil War?

E Pluribus Unum—one out of many—places incredible stresses on the human heart.

Our inheritance of political and cultural freedom, always implying tolerance for the stranger, becomes a paradox: our very democratic freedoms may breed a fear of the enemy.

Fear of the "enemy" has sometimes overwhelmed my People of Peace.

The purity of the early peacemakers, say John Woolman and William Penn, who centered their peaceful efforts on a personal belief in a loving God, has been lost. The Abolitionist Movement for freeing the slaves began in non-violence. Yet Frederick Douglass, John Brown, demanded violence and even William Lloyd Garrison acquiesced in its use as the way to free the slave, the south the great "enemy." As John Brown said just before he was hung for leading the insurrection in Harper's Ferry in 1859: *the crimes of this guilty land will never be purged away but by Blood.*

Even our major peace movements so often abandon their principles at the advent of war, ready to defend the nation against its enemies.

In 1848, after the end of the Mexican-American War, Elihu Burritt founded the League of Universal Brotherhood. Its goal was to obtain the support of the American working man against war. The first secular organization in the world concerned with peace, Burritt accumulated around 50,000 signatures to the first peace pledge:

> *Believing all war to be inconsistent with the spirit of Christianity, and destructive to the best interests of mankind, I do hereby pledge myself never to enlist or enter in to any army or navy, or to yield any voluntary support or sanction to the preparation or prosecution of any war....*

In 1861, thirteen years later, the United States fought its bloodiest war between its own citizens, many of whom must have signed this pledge.

Before World War One it was believed that the socialists of the western world would never fight each other in a war. From1914 to 1918 those same socialists killed each other in the most murderous war of all times until World War Two.

Before World War Two college students in the United States and in England banded together to sign the Oxford Pledge, promising never to "...fight for...King and country." On April 12, 1935, 60,000 students around the United States went on strike against war. In November of that year 20,000 marched in the streets. Thirty-five hundred students booed the president of the City College of New York when he objected to the reading of the Oxford Pledge.

From 1939 to 1945 these students were the cannon fodder of World War Two. Many volunteered to fight.

People so easily forget their promises to peace when the dark shadow of the "enemy" threatens their nation. War and the destruction of a nation's enemies apparently demand a higher loyalty than any promises to peace.

The difficulty, of course, lies in the paradox raised by Justice Holmes, supported so vociferously by Bela Tiffany: "society has rested on the death of men." Sometimes, men must kill as there are real enemies. Nationally, the slave owners who preferred to splinter the Union before giving up their right to own another human being; internationally, the Nazis who killed Jews, their enemy; the Communists who opened the Gulag for their own dissidents; domestically, the need for self-defense, and, even more wrenching, protection of the innocent from the attack of terrorists.

Is not the greatest wisdom being able to discern real enemies from false? Is that not the wisdom possessed by the most astute People of Peace? Henry Thoreau and Levi Tiffany, who understood our war with Mexico in 1846 was a war for Empire, self-interest, not for justice; Jonathan Woolman, who understood that slaves were not merely a piece of property but real human beings with rights of their own; Ethan Allen Hitchcock, who understood that the American Indian was no Devil Incarnate but a member of the human family; our efforts today to separate the terrorists from mainstream Muslims.

Yet we must be most careful about drawing the conclusion that "enemy-making" is an uniquely American characteristic.

The Holocaust in Germany, the Gulag in the USSR, slaughters in Cambodia, Rwanda, Sierra Leone, Nigeria, the Congo, Central America, and the former Yugoslavia, the wars between Iran and Iraq, Ethiopia and Eritrea, and the terrorists who loathe us so much are all characteristics of this awful century, rooted in the fear of enemies, national, political, ethnic, and linked to the awesome quality of human weapons.

Sometimes not even weapons sophisticated in nature: the axe, the machete, the sword, arms and hands viciously chopped off…retaliation for what? No one is ever quite sure of the cause once the mutilations cease.

Are such horrors only unique to the twentieth century?

Do they have far deeper roots, stretching back beyond civilization's beginnings?

Did primitive societies, long before western civilization, possess similar qualities of enemy-making and barbaric use of weapons?

A recently published book, *War Before Civilization,* by Lawrence H. Keeley, reaches the conclusion that many primitive cultures were highly war-like, with deaths and woundings from war far in excess of those of western civilization, even its most recent wars.

Was there ever a Paradise of primitive cultures where man to man was beyond our violence, living out life in peace, without enemies, beyond the weapons of our fathers?

In answer, Levi D. Tiffany brings out his *Journals* for the years before the Civil War.

He leafs through them, page after page.

I searched for that Paradise for my whole life. Found it in the tenderness of which Bela and Bill Green murmured, in the natural world in which I lived, in my Bible, the words of Christ in the Sermon on the Mount, the voices of all those American peace-makers. I wrote of their words in my Journals *for so many years, the Paradise formed for a few brief moments in an America before our wars, before we turned our brother into our enemy.*

What happens to that Paradise, the one for which all humans long? Why does it so easily disappear? Why is it pummeled by war and violence, sometimes crushed to nothing? Is that capacity for violence an essential part of the human soul?

You, we, must speak of those ancient cycles of war and peace, violence and compassion if we are to ever understand those men and women of so long ago, of cultures before there were cities, before there were towns, before western civilization. Were they violent? Were they peaceful as so many believe?

In your Journals, *kept over so many years, you once searched those cycles of the past for the common stuff of human nature: were men and women in primitive times the same as we are today? Did they differ in the way they made enemies, in the way they used weapons? Is there some common human quality to our need for enemies that leaps over time?*

Search yourself, your past, your own Journals *for the answer to these questions which so haunt us all.*

And which must be answered if we are ever to reach a world beyond the weapons of our fathers, our elusive dream of a world at peace.

Once on a writing fellowship at the Wurlitzer Foundation in Taos, New Mexico so many years ago, I sank deep into the paradox of those questions Levi asked. I hiked for days through the high mountain desert of Northern New Mexico, searching ancient Anasazi ruins, one of the oldest Native American cities in the country, for some answer to the puzzle: were these ancient societies the utopias so many writers claimed, or were they no different from ours, sometimes peaceful, sometimes warlike, fearful of the enemies about them, well-armed with the weapons of their day?

My *Journals* record that search.

23 August, 1987
Seated in a cave
in the fifteenth century
Anasazi ruin of
Tsankawi.

Not five miles from Los Alamos, where the atomic bomb was invented and produced for use in World War Two, Tsankawi is one of the fourteenth and fifteenth century abandoned Anasazi Pueblos in Bandelier National Monument in Northern New Mexico, I've explored its caves and pictographs and touched its pottery shards, scattered over this bone-dry, hardened earth. Now, I stretch out in this dark cave, eat my lunch, work on my Journal.

Earlier, I parked in the deserted gravel pit below the Mesa where these ruins lie. The high desert sun scalded my eyes. I fumbled for my dark glasses, grabbed my rucksack from the backseat of the car. (I always carry raincoat, water, lots of it, a snakebite kit, and food when hiking in the desert.) Slipping into the pack's worn and comfortable straps, I followed an ancient Anasazi path, worn deep into the soft, white stone.

To my left as I climbed toward the mesa top was another path cut as deeply into the stone. On sudden impulse I followed it. Where it ended, I saw a group of cave openings. I stooped and peered into one. The sun at my back blinded me. I took off my dark glasses. The interior of the cave slowly appeared: smoke-blackened walls of five hundred years ago, a rough petroglyph etched deep into the wall. Carefully, I edged into the cave, wary of rattlesnakes that might be attracted to the cool stone floor. When certain the cave was empty, I entered, sat, turned toward the cave opening.

The view exploded before me. Piñon trees, sage, earth, bone-white clay spread across the canyon floor. Silence hymned in my ears: the silence of Chartres Cathedral in France where once I had worshiped, the silence of a Quaker Meeting when it works, silence resonating, buzzing in this hot and golden afternoon. Around me lay the unexcavated ruin of Tsankawi, the ancient home of the Anasazi, the "vanished people" of New Mexico, deserted for over five hundred years, The cliff walls were covered with pictographs and petroglyphs, ancient trails worn deep into red and white sandstone, while pottery shards sprawled over its mounds and tumbled walls.

I do not know how long I sat.

I longed to know the past of the Anasazi as it really was, not as the west romanticized it, not as those writers I read had imagined it, but what it had really been like. Was it true that it was less violent than our time? Did the writers I read who made it into a Paradise, a utopia, really speak the truth? Or had they romanticized the past in a desperate search for a perfect place to balance out our modern brutality, our incessant search for enemies linked to our development of ever more obscene weapons, the kind invented in Los Alamos, less than 5 miles away? Was it that the violence of the Anasazi only had a different quality, harder for us to imagine?

I shut my eyes and thought of the Anasazi as they must have been in the night. I thought of the Anasazi when fever or accident struck. Few medicines to break the fever, no splints to heal the broken bones. Most of all I sensed the darkness, the night black and thick and heavy about them, the stars pressed down in a bright glitter upon the earth. Deer bleated a dying cry as wolves howled a victory cry.

In the blackness of the night in which the Anasazi lived I sensed a violence different from ours, mysterious, rising from the night, rising from nature, rising from drought. Sometimes not enough food, nature threatening with unexplained illnesses, inexplicable accidents, inexplicable rages, the violence of a thunderstorm, lightning streaks stabbing at the earth, even, sometimes, momentary apogees of war over territory or scarce food and water. Violence outside our modern technology, violence formed from both nature and from man, violence only propitiated by magic...but violence here as well, humanity naked to its force.

What enemies did the ancient Anasazi fear?

What happened in times of drought where there was not enough food? Starvation ruled? Enemies attacked?

Did clan fight clan for food? Tribe fight tribe? Family against family? Use the club, obsidian dagger, bow and arrow to kill?

Why were these Pueblos built on mesas or hidden high in cliff walls, adobe homes and villages that could be so easily protected from an enemy?

I squeezed my eyes even tighter.

Did I hear those ancient cries of warning tumbling through the ages?

Men shouting their obscenities as they defended the narrow path to their homes?

Further up the trail I know of another petroglyph, etched into the canyon wall, a human face, locked into horror and surprise. Most simple lines form a frightening impact: the ghost of enemies struts within these canyon walls.

The Anasazi lived in nature and worshiped it. Yet, by the strangest paradox, that nature was also the greatest of their enemies: illness, death, drought, cycles of drought repeated through the generations.

They believed they could influence that natural force with magic, rites and rituals.

But what of human enemies?

Perhaps the Anasazi experienced long periods of tranquility when climate, food, and community balanced each other. But when rains ceased and famine prevailed, did clan fight clan and family battle family for food? Is that why the great cities of the Anasazi to the west were abandoned? Why these villages were hidden on high mesas or in cracks on cliffs? Were the primitive weapons of bow and arrow, knife and stone used in mortal combat?

Did they kill face to face, blood hot as it spurted over fingers?

Were these really the homes of a people peaceful for all times as so many wish to believe today?

Or is it that our writers must create utopias out of the past, imagine places and times without violence, antidotes to the terrible violence of our day?

Does it make any difference that their weapons—bows and arrows, knives and stones—were far less dangerous than our weapons of today? They could still kill. An arrow could penetrate the heart, a stone could smash the skull and an obsidian knife could slash the gut.

I do not know the answer to these questions I ask as I listen with my inner ear for the night of the Anasazi.

But I do understand that it is so easy to worship our past as something it never was. So easy to deny our primitive need for enemies and all the attributes that accompany them.

So easy to deny the deadly quality of primitive weapons.

"Enemy making" must have been part of the Anasazi world.

"Enemy making" must be common to human nature: primitive cultures, sophisticated cultures, western nations, eastern nations share this quality. Is this our fundamental human dilemma?

Five miles away from where I sit contemplating this past, men in air-conditioned laboratories live comfortably with the world of enemies. They design weapons that, to the Anasazi, would have been incomprehensible, destruction beyond imagination.

How did Northern New Mexico get from the primitive weapons of the Anasazi to the technological horror the scientist creates in Los Alamos? From bows and arrows, stones and obsidian knives to the atomic bomb?

Killing carelessly, killing at a distance? The real mark of western civilization.

Killing without experiencing the victim's pain?

The Third Chorus

In the Barn at Blandford, Massachusetts

William Green
Bela Tiffany
Lydia Tiffany
Levi Tiffany

William Green comes out of the darkness in the Tiffany barn, confronts Levi, Bela, and Lydia.

But it just ain't enough to talk of enemies. It's weapons we got to understand.

Bela

Weapons?

William Green

Yeah, weapons. You and me, Bela, we fought with that ol' .58 caliber rifle, remember? Sweatin' with fear. Rammi' the minie ball down the barrel with tremblin' fingers, scairt to death we'd get shot afore we could fire again.

This boy here, why fifty years later, his Daddy flew airplanes.

Bela

And he done had that M-1 Garand, fired eight rounds without reloading.

William Green

Now they put all that money into bombs that destroy whole cities.

Lydia

Missiles that fly them through the air. Kill women and children.

Bela

Missiles they think'll kill other missiles.

William Green, so excited the old man coats his words with spit

We fought fair. Why this ol' bluebelly and me had to stand up on the battlefield, face each other at a hundred yards. Remember, Bela? Courage? Honor? Women and children safe?

Now, their men don't risk nawthing!

His words expectorate from his mouth, contempt now their coat.

Nawthing!

Their president even brags "no casualties." They win a war without risk. They don't fight. They bomb from 15,000 feet!

Bela

They spend all their time developing those great weapons that kill from a distance. A man don't even know he kills. Why, I saw the eyes of the men I kilt. Felt closer to them than anyone I knew. It were awful.

To kill so close.

A man don't want to kill after he sees a man he shot die. Why that man is the same as him. Got kids. A family. Maybe a little farm. Like you, Bill Green.

William Green

But we fought fair, Bela. You had your chance. I had mine. And, finally, one of them got me.

Bela

Shot me, too. Just the way they shot the boy.

William Green

That's what your generation gone and done, boy. That's what it gone and done.

Spend all that money, that "Research and Development" you call it, on weapons that kill from a distance. Ain't got the guts no more to kill a man face to face in war.

Kill women and children from 15,000 feet.

What kind of men are you?

Soldiers!

Hell, you ain't soldiers.

You're God damned ('scuse me, m'am) cowards!

Bela

Murderers!

Lydia

Not just those weapons but those pistols, machine guns, automatic weapons that spray bullets everywhere, that anyone can use to kill.

Bela

Plastique, bombs anyone can make.

Lydia

Selling those weapons to poor countries when they should use their money for food.

William Green

It's madness, boy. Your world done gone mad.

Bela

Do you think the framers of the Constitution woulda' put in the Second Amendment if these new weapons had been invented?

Why, our rifles had scarcely changed from the Brown Bess of the Revolution. Still single shot weapons.

Would Thomas Jefferson approved some damn fool kid....

William Green

Or grown idiot....

Bela

...be allowed to own a gun that sprays hundreds of rounds a minute?

Solomon Lombard calls from the far distance.

Nonsense! We didn't want no militia or no person carryin' weapons like that. Wouldn't allowed it in the Constitution.

William Green

Wouldn't 'llowed no atom bombs neither. That's what you got to tell us, about the madness, about your journey to that place of madness, Los Alamos where they invented the atomic bomb.

Lydia

The madness of your time. Little children killing little children with automatic weapons. Great bombers killing at a distance while ordinary folks go about their business, not even knowing their country is a killing innocents.

Levi

Your story's not just made from enemies. It's made of the weapons you invent to kill them, how they grow in obscenity each year, that's the tale that must be told.

William Green, Bela, and Lydia in consort

To speak of enemies is but half a tale. To tell of weapons completes the horror of the world you and your generations have made.

CHAPTER FIVE

✯ ✯ ✯

The Persistent Quality of Weapons

Better Ways of Killing Each Year ... With a Closing Chorus

America is awash in weapons.

More atomic bombs than any other nation. More missiles.

More airplanes.

More guns.

Who really knows how many?

Estimates range from 200 million to 250 million guns of all kinds owned by Americans, a little less than the population.

The number of handguns increased in almost geometric proportion during the civil disorder in the 1960s, an estimated 16 million in 1960, 27 million in 1970. By 1997 there were 67 million handguns in the United States. After September 11th, there was a sharp increase in the purchase of weapons.

The murderous capacity of these weapons has simply exploded, evolving from the old fashioned revolver with six shots to the pistol that can spray a clip of ten bullets in seconds...not to mention the assault weapon, even deadlier.

Just as makers of atomic weapons and missiles pour money into research and development, so do those corporations that produce handguns, making weapons and bullets more lethal each year.

The federal government plays its role in support: it buys and exports massive numbers of weapons, dominating Europe for first place in the international arms market.

The military plays its part: it buys weapons of mass destruction while the handguns it adopts for use become the model for new domestic weapons.

The media play their part: they show us how attractive weapons are, so easily destroying the dangerous enemy in the movie and on TV, the image of our darkest dreams.

The legislatures and executives of State and local government play their part: they refuse to pass and to administer legislation concerned with controlling the use of weapons.

The National Rifle Association plays its part: it continues to sell that old myth, "a people well-armed and dangerous," while the Brown Bess actually was a weapon of startling inaccuracy, the weapons of the antebellum south most dangerous to the Native American and to African American slaves.

Thus the weapons industry of either weapons of mass destruction or hand held weapons has an enormous support system and, so, has never had any stringent controls placed upon it, perhaps one of the most untouchable industries in the United States.

The assumption of the United States—people and government—appears to be that the production of weapons, the research and development of weapons, is an activity intended for the public good.

Each generation must be given ever more deadly weapons to face and defeat their "new" enemies.

A delicate ballet is choreographed by the weapons industry, the media, the government, corporate America, and the NRA: a new enemy is discovered each generation requiring the development of more virulent weapons to defeat that enemy.

As *Beyond the Weapons of Our Fathers* shows in the first chapter, this ballet has been danced since the discovery of this continent and the founding of this nation.

If there is one thing I have learned over the years of searching for the causes of our violence and our wars, it is that men must improve their weapons of destruction each generation. In some deep way, seldom studied, seldom examined, the dark fear of our enemy links with our technical skill at invention. Each generation weapons evolve in killing power, evolve from gunpowder to atomic force, then to biological and germ warfare, missiles for delivery always available.

Killing today in western society differs so from the weapons of Homer's *Iliad:*

> *... and [the spear] split the archer's nose between the eyes—*
> *it cracked his glistening teeth, the tough bronze*
> *cut off the tongue at the roots, smashed the jaw*
> *and the point came ripping out beneath the chin.*

Close combat in the Civil War experienced by Bela Bentley and William Bigbee Green shared this terrible quality of hot blood and bare steel. Once in France I fixed a bayonet on the end of my Garand Rifle. I understood the deepest fear and terror: to kill face to face so differs from killing at a distance.

Visiting the city of Los Alamos where the atomic bomb was perfected, after meditating in my cave in Tsankawi, scarred me with the knowledge of that difference in killing, now we kill at a distance, then face to face, perhaps the mark of our generation.

25 August 1987
Los Alamos, New Mexico
My Journal
Near Fuller Lodge

Los Alamos, New Mexico lies not a ten minute drive from Tsankawi. Los Alamos, where the atomic bomb was conceived, built, and shipped, first to White Sands, New Mexico for its test in July 1944, then its use in Hiroshima and Nagasaki, Japan.

I knew I must visit this place of death if I am ever to understand our propensity for violence, our weapons, our destruction of enemies at great distances without suffering ourselves.

The experience this morning shattered me.

The longer I stayed in Los Alamos, the deeper I penetrated into the paradox of our era: humanity now technologically capable of eradicating all life on earth in violence, an accident, an act of terrorism, or war. As I drove on the highway from Tsankawi, my heart thudded hollowly. This twisting, high-desert road, the one the bombs had first followed on their test to White Sands in southern New Mexico and, then, their delivery to the South Pacific in World War Two: Little Boy to Hiroshima and Fat Man to Nagasaki. I had passed ancient Pueblos, constructed in the fifteenth and sixteenth Centuries, on the drive from Taos, Otowi Junction with its 1920 Bridge crossing the Rio Grande River, the turn off to Bandelier Monument with the historic Anasazi ruins of Tyuoni, Tsanswaki, Navawi.

In Los Alamos two museums memorialized the discovery, design, and use of The Bomb. One, sponsored by the local historical society, centered largely on the life style of those who lived in Los Alamos during World War Two and invented and built The Bomb. The other, owned and managed by the federal government, celebrated the historical background and the technical process of The Bomb's development.

I stumbled on the government museum first, really a modern suburban office building in style. I parked, nodded as pleasantly as I could to the guard at the entrance, then stepped inside the building. Memorabilia of The Bomb were spread over a long wall, between statues of General Groves who controlled the military development of The Bomb and J. Robert Oppenheimer who directed the scientific effort.

A video player was embedded in the wall between their statues. Selection could be made among tapes about The Bomb, its discovery, its use.

For an hour? Two? Three? I was mesmerized by those trembling images in black and white recalling the era of World War Two, the war in which I was wounded at nineteen, movies of both Groves and Oppenheimer, the scientists who helped them, the politicians who controlled them.

Flickering shadows recalled the Nazis, the Holocaust, Pearl Harbor, American Japanese imprisoned in our concentration camps, and the explosion of the first bomb, Trinity, in White Sands, 200 miles south of where I now sat. Satiated and stunned by the memory of that war, the horror of its butchery, culminating in the shock of atomic annihilation and the discovery of the camps of the Holocaust, my heart thudding again. I lunged from these movies and stumbled toward the exhibits.

Around the corner full scale models of Little Boy and Fat Man were displayed. My heart seemed to pause for an instant as I reached out and stroked those grey-metalled, dull-shining surfaces: originals of these machines had killed thousands of people, destroyed cities, and ushered in a new relation between man and nature.

The men who dropped them never saw the horror of their act and, as one said, never had trouble sleeping.

They were the enemy, weren't they? The Bestial Jap?

In sudden nausea I stepped back from the bombs, out the door into the parking lot, tugging my collar for air along the way. The guard stared peculiarly at me. I wondered as I rushed by him how many visitors left the museum as I did, sickened by what they had seen, technology displayed as obscenity.

The streets I drove glorified our capacity for destruction, the main street even named "Trinity" after the first atomic bomb, another called "Eniwetock" after the atoll in the South Pacific where the hydrogen bomb was first exploded.

I could not believe we would so honor weapons of such obscenity by naming streets after them.

In fact Los Alamos seemed to reek of some poisonous atmosphere. I could sense it in my bones, the memorialization of machines of death. Instead of mourning the death of 200,000 persons and the destruction of two great cities, the monuments glorified the fact of killing.

Even if we had to invent and use The Bomb, what was wrong with my country that it should so glorify those acts of brutal killing at a distance?

Just off Trinity Street, I turned off the road to Fuller Lodge, now a community center, once the heart of Los Alamos during the war. The guesthouse behind the Lodge had been turned into a historical museum. I wandered through its models, photographs, and artifacts, memories of what life had been like in Los Alamos from 1943 to 1945. One fact resonated in my mind: the average age of the men who discovered and built The Bomb was the late twenties. At that age patriotism so easily subsumes all other values: there is great patriotism, love for country, linked with intense fear of the enemy, in this case the sub–human Jap. Joined to these values of patriotism and fear is an immense worship of and skill in the development of weapons technology and little understanding of its disastrous consequences. Even their leader-director, Oppenheimer, had just turned forty, still a young man.

Across from Fuller Lodge lay the pond that had separated the residential areas of the town from the buildings where the bomb was designed and constructed. Though the buildings had been torn down long before, by careful pacing and use of the map I had bought in the museum, I located the site of "Gamma Building," where much of the work on The Bomb occurred.

My horror increased geometrically when I stopped at what was the site of the front door of the building. For an instant I lived in both past and present simultaneously. Forty-four years before the inventors of The Bomb planned and plotted exactly where I stood. I closed my eyes. In my imagination I saw young men rush by me into the door as they struggled to meet the deadline imposed by the Allies' demand for unconditional surrender. They glowed with satisfaction as they solved the problem of design; trembled with uncertainty as they neared the first test at Trinity; celebrated with fierce delight when the bombs were exploded over Japan.

I turned and fled, recorded my thoughts and feelings in this *Journal* with a trembling hand.

29 August 1987
Seated at MacDonald's
Los Alamos, New Mexico

One last journey to this City of Death.

One last search into the memories of our past of violence and of enemies.

One last effort to understand the role of weapons of destruction in our modern world.

The men led by General Groves and J. Robert Oppenheimer, inventors and builders of The Bomb, were rooted in far different traditions and beliefs about the nature of reality than the Anasazi, those ancient ones who lived in this same high desert. Human beings in the universe of modern scientists are separate from the natural world, rational beings first of all. Immutable laws, not Gods, not magic, control the actions of chemical, physical and biological systems, which, working together, create our universe. The rational mind through hypotheses and experiment can understand those laws and use that knowledge to change the world.

Men do not need to propitiate nature with magic: they can rule it.

This knowledge and control would inevitably bring about improvements in the "public good." Scientists of the late nineteenth and early twentieth centuries, giants who made possible The Atomic Bomb of the 1940s—Albert Einstein, Ernest Rutherford, Niels Bohr, Leo Szilard, Enrico Fermi—believed that their discoveries would improve life for all people. Progress, the diminution of evil—human suffering and pain, illness, poverty—was the inevitable product of the application of the laws governing the behavior of the universe.

In World Wars One and Two that belief was irrevocably shattered. Science and Industry, serving State and Business, developed the machine gun, the tank, then poison gas, the long range bomber and The Bomb. Wars became agonized attritions, killing done at immense distances, fought until the enemy, soldier and civilian, lay prostrate and smashed: the tradition of unconditional surrender.

Violence, fear of the enemy, linked with corporate, economic, bureaucratic power and obscene technologies of death became the seamless edge of a dark blanket thrown over the globe, a blanket woven from our knowledge, technology and skills.

We have come to believe that we can totally manipulate our reality through science and its inventions. We can destroy cities to win a war. We can alter life through bio-genetics so as to "improve" it. We can kill so as to give freedom to the killed.

Our weapons, The Bomb most of all, provide us with the stunning proof of our Godhead while, simultaneously, giving us total power over our enemies, whether real or not, able to destroy them at great distances, never seeing the destruction and the consequences of our acts of violence.

This power of total destruction at a distance fascinates us. Not just with The Bomb but with our lesser weapons as well. We enjoy tinkering with the technology of death: cleaning and polishing a rifle, "zeroing" it in on a rifle range, lazy comradeship after firing it with friends over a beer, comparing our prowess.

Oh, we deny it so much, our enjoyment of the moment of death when the bullet, the bomb smash our enemy, the knife, the bayonet slice his flesh. Though we can never admit our pleasure, at some place deep within our souls, we delight in that act of destruction, total power over another, else why do we repeat it so many millions of times?

✯ ✯ ✯

The nightmare of those weapons of my forefathers, aimed at my soft parts, returns and, with it, Peter Zachary who rests easy on his rifle as he stares at me

> *Those weapons, why ain't they wonderful? So much better'n this 'un.* [He shakes his rifle.] *They make America the most powerful nation in the world. What would our country be without those arms? Those bombs? Planes? Missiles?*

Humphrey Tiffany

> *Mock those guns? That power? Why, those weapons keep the savages from our shores. Let us enjoy our wealth and goods.*

The Lombards

> *Our money.*

Peter Zachary gloats

> *That's what your People of Peace never understand. Without those weapons, without improvin' 'em every year, all those God damned colored folks'd be a takin' everything we got. Our wealth. Our women. Our weapons protect us, give us strength. Member what yo' Daddy said: "Without a gun a man is nothin'."*

William Bigbee Green

> *But Grandpa! We got to learn another way. Weapons ain't like your Kentucky Rifle no more. Or mine and Bela's .58 caliber rifle. They got these planes. These super bombs. Killin' women and children at a distance. We got to change!*

Zachary, Humphrey, Consider, the Lombards in chorus

> *Ain't you ever gonna learn? We can't stop. We started down this path when the first man picked up a club and bashed the head of the man who tried to take his woman. Then that man's son grabbed a rock, hid in the bushes, threw it at the man with a club, busted his skull, took the man's daughter for his own. Goes on forever and ever. You oughta jus' be plum'*

thankful you live in a country where your leaders spend so much on weapons. Let us run the world.

My nightmare of the weapons of my forefathers trained at my soft parts beats at me with even greater force. Is this all, then, there is? Enemies and weapons recycling endlessly through time? Men determined to blow up the world so as to destroy their enemies? Men killing at a distance, free of shame over the victim's pain?

Levi Tiffany, his old eyes wet with pain

We all feel such despair at times. But we can't quit. All that lies between life and death is those who say: "No, No, I won't do that. Kill. Maim. Hurt. I won't use cruel words, deny opportunities to those weaker than I."

They form the real America, not the men of war.

Bela Tiffany and William Green

Courage to live by what you believe.

Levi

It's the search for understanding that so matters, the search I made in my life with my Journals, *the search you have made in yours, a life beyond the weapons of our fathers. In that search lies the hope of the nation, beyond the voices and the acts of men who believe in war and violence, in weapons to destroy their enemies.*

William Green

Even beyond the effort to understand. Maybe that's your problem, the problem of your time. Spend so much seekin' answers, reasons for everything. Maybe there jus' ain't no big answer but a lot a little answers rolled into one. Jus' people bein' themselves for centuries.

I found my way by not lookin' for answers.

On my farm.

Out in nature.

Not with men but in silence.

No damn fools a talkin' too much.

Jus' sky and clouds and the plow a cuttin' through the soil and fishin' once in a while, a lyin' in the sun, safe, free…it took me years to get over the war and feel free again for, boy, I did some awful things, never yet forgiven myself!

Bela

I found my way beyond the killing in beauty. First time I ever touched a camera was after the war. Gave me such peace. going out into nature, taking photographs. Posing men and women in my studio. Seeing how much my work pleasured them. Making something beautiful out of nothing, why this saved my life after the madness of the war.

Levi

I found myself in my Journals, *daily recording events outside, discoveries within.*

That's the same thing you do, constant effort to understand.

Your Journals *are your own record of your search for peace, kept for over fifty years.*

Go back to them. Read them. In them you tell of your search for peace. In them you discover the ethic for which you search, formed and made from your own efforts over a lifetime.

Your Journals *contain the truth you so long to hear.*

In them you will discover your life beyond the weapons of your fathers.

Part Three —

Beyond the Weapons of Our Fathers

✯ ✯ ✯

And, so, it is deep within my own heart
and my Journals over decades
that I must discover the answers I so seek…

CHAPTER SIX

✫ ✫ ✫

Discoveries of a Life Beyond the Weapons of Our Fathers

Found in Forefathers, People of Peace and Recorded in My Journals

I have kept Journals for all my life, sometimes failing miserably, sometimes working on them for long periods. This discipline started after I was wounded, as if the only place I could state all I really felt was in my daily effort to use words, turn them into sentences, paragraphs, pages, learning what went on within and without through the words I wrote.

I do believe that effort saved my life.

I did not realize for so many years that the search I made was to come to terms with the violence in which I lived, reach out toward ways of peace, record moments of peace when they came, their content, shape, and form, deal with my own shame.

Out of that struggle I have formed an ethic for living in this violent world, beyond the weapons of my fathers.

In so doing I have learned that there are few guides for those who seek to create such an ethic. Our society is not particularly interested in pointing out ways of living beyond violence. We are trained to "win," to use the weapons of the day to defeat our enemy. Discovering these new ways—for me, living in nature without exploitation, efforts at compassion, a search for beauty—are recorded in these Journals of over fifty years.

I turn now to that lifetime of records in amazement as those moments of discovery rise to the surface.

I also search that past in an unexpected awe…it strikes me so odd that three generations later I follow the same path as Levi Tiffany: using words to make sense of life, seeking ways of peace beyond the daily rut, developing an ethic for living in this violent world.

Is there really no time?

Do those alive, those dead, link in ways we seldom realize, our mighty effort to understand the universe and our role within it?

7 September 1994
Colorado River
Near Kremmling

I was wounded fifty years ago today in France. I always drive to the Rocky Mountains from Denver so as to spend the anniversary of that event somewhere in the natural world, often fly fishing for trout.

The car is parked about a mile away on the banks of the Colorado River. The walk to the place I fish follows the river, far enough from parking lot and campground to find some isolation. The Colorado here is about 50 yards wide, perhaps 6 to 8 feet deep in the middle. I wade the edge of the bank, never much over my thigh as the current could easily knock me over. Most of the smaller browns and rainbows—up to 15 inches—are near shore anyway, lying in calmer water where they expend less energy as they wait for food to drift down the current.

At this time of year I usually fish with a Royal Humpy, hook size 12 to 14, red or yellow patterns. I cast upstream, let the fly drift backwards. Sometimes it simply disappears in a great suck as a trout takes it in. Sometimes the fish, particularly if a rainbow, will rise straight into the air in a curving leap, an arch of beauty.

The curve of a trout as it takes a dry fly and leaps into the air to shake the fly from its mouth is one of the finest sights I know.

Especially since I now fish with barbless hooks and never kill a fish unless, somehow, it's been hurt so badly it won't live.

Just the beauty of it.

In all the crises of my mature life, reactions to moments of stress and violence, I have come to the natural world for cure. My father took me into the great outdoors fishing from the time I was five or so just as his father had taken him. After I was wounded, Dad took me on a two week canoe trip where we fished and camped and spoke of little but knew each other in real companionship. Later, when my life really collapsed as a consequence of that wound, I moved to rural Connecticut, became a gardener, survived by living in the out of doors, my body supple from work with plants and earth. When my life fell apart again in the 1970s after a divorce and my flight from the merciless system of the Massachusetts Institute of Technology and Washington, D. C., where I worked, once more I returned to the natural world, living as simply as I could in a small house by the Canal on Cape Cod, again earning my living as a gardener.

This natural world in which I worked was all mine.

Not because I owned it. Because I couldn't own it.

Because it is free.

In its freedom it is mine. And I am its.

For me, at least, the natural world rests me from anxiety, frees me from the fear of enemies. I sense in it the evolutionary force that reaches back through time and marks that moment when life was formed. I sense in it the hope of resurrection, birth and life and death rising from each other. I sense in it the source of all beauty, colors and chords, from which men and women derive their works of art.

12 March 1993
A Walk in
Matala, Crete

Until this morning it has been terribly cold, so cold all we do is go to bed after dinner as the hotel room is unheated. But, slowly, slowly in the morning the air warms and, after we have had our breakfast of fruit and bread down near the harbor, we walk down the back road and through the hills of Crete.

Today—I'm writing this in bed, blankets wrapped around my shoulders, my fingers icy—we discovered the partially excavated ruins of what had once been a Minoan port city, perhaps in 1800 B.C., the same period as Knossos? Great pots 5 and 6 feet high were scattered in the excavated ditches.

We walked up a draw toward the hills. (I think the valley once must have held a stream.) And on the ground were mounds of pottery shards. Great red pieces, handles of pots as large as those we had seen below. I cannot even imagine how many pots there must have been to leave so many remnants.

At the top of the draw we turned south toward the shore and the path rose higher until it reached a great plateau.

At the edge of the plateau cliffs, 200 or 300 feet high, fell straight down to the blue sea.

Not a house.

Not a person.

The sea rising, falling, sending white capped waves to beat on the sandy shore beneath the cliffs. The air pure and clear. In the distance the smoky haze of Africa.

And in the hills around us lay bands of fossil shells, clams or oysters. We could break them off and hold them, in one hand the shells of a billion years ago, in the other the shards of a lost civilization.

Time and nature seem to be the same, gift of eternity.

Yes: in the natural world lies a place that rests us from our anxiety, the enemies of our modern world.

Not that nature is benign.

Far from it.

Spend the afternoon caught on a Colorado mountain in a thunderstorm with lightning snapping the air, filling it with the sharp smell of ozone, the ground reverberating from the smashing roll of thunder. Be driven to the lee shore while sailing through a storm, the boat rising, falling, slamming into the rocks below until it is smashed to pieces. Ride a canoe down through roaring rapids, knowing that any rock can tear out the bottom.

Nature has a violence all its own, magnificent, austere, yet far above us.

Be trapped in a flash flood camping in a Rocky Mountain canyon. Lightning strikes so near, the black shutters of the night momentarily flash open with streaks of yellow and red; the smell of ozone from a close strike stuns the air; the tent trembles from thunder.

Nature differs so from the malevolence of humanity. Human violence is directed at another as "The Enemy" from an act of the will, centered on destruction from desire. Natural violence has no such intent, totally indifferent to one's fate, never willing any particular result. Lightning is not aimed specifically at you, the shore does not reach out to smash your boat, the rock does not leap out of its way to attack your canoe, the flood rolls toward your tent and all others, not caring who is hurt or killed.

The dark rolling water of this river before me, ripples sliding over just visible red-brown boulders, the rising hill across the way. All these, then, contain a power I so love: majestic, intimidating, graceful, always poised in violence, yet always present with beauty, a force I cannot understand, yet one which I must call the Divine, beyond the world of enemies and weapons.

15 October 1991
Benedictine Monastery
Christ in the Desert
Abiquiu, New Mexico

My search for a life beyond the weapons of our fathers has led me to new places I might never have experienced. In 1984 I lived in Chartres near the Cathedral where I completed the first draft of my memoir, *On Being Wounded*, built around my decision to give up my father's weapons. In Chartres I first clarified my understanding of the spirit of compassion as a core value of my People of Peace. Here, now, at this Benedictine Monastery, set in this high desert with its lavender and purple mesas, its dark green pines and blue-green sage, its stillness, the dark Chama River flowing silently through its meadows, the ancient pueblos of the Anasazi high above on those mesas, here in this adobe chapel, I merge again with that same spirit.

I must explore this compassion and what it means to live with integrity in this world of violence, killing, enemies, weapons, and death. I must explore it in this high-desert beauty, in this place of ancient cities of the Anasazi, oldest in my land, in this place where the atomic bomb was invented and tested, the most wicked weapon of mankind.

How does this monastery, dedicated to compassion, exist here in the desert, Los Alamos where the bomb was born and tested, not 60 miles away?

Is this not the true place to discover the meaning of a life beyond the weapons of our fathers?

17 October 1991
Benedictine Monastery
Christ in the Desert
Monk's Cell
Abiquiu, New Mexico

In this silence, this barren cell with a kerosene lamp, a bunk, a wood stove, blankets, a chair; in this land, its red and lavender mesas, its sage, piñons, its bone-white clay, its link to eternity; a place to continue my search for a way of life beyond the weapons of our fathers, outside a society dedicated to the destruction of its enemies, rooted in such fearful anxiety.

My book, *On Being Wounded*, has been published.

It only opens the door to what has become a life-long pursuit: struggling with my anxiety, the enemies it forms, why the nation acts as it does, searching for an ethic by which I can live.

Later
Common Room
Monastery

Food is eaten silently here. A Brother reads from a religious work during the main meal. We, the Brothers and their guests, eat in silent companionship, relieved at not having to make small talk.

The monastery lies at the end of a twisting, clay road, treacherous when wet, forty-five minutes from hard top. It nestles beneath great pink and lilac and lavender mesas, its buildings so close to earth they seem to have grown from it. Its adobe church, guest houses and convento look as though they have been here for hundreds of years.

I worship in the chapel with the Brothers five or six times a day, often beginning at four in the morning. Worship is simple, the traditional offices of the Benedictine service, Gregorian chants welcoming the divine into our day at dawn, thanks for food at dusk.

What do I seek as I worship? What in sweet God's name?

A place where there are no enemies, no weapons, no anxiety, a place where trust abides, a place where gentle words and kind thoughts rest us from our pain.

With Christ:
Again, ye have heard that it hath been said, Thou shalt love thy neighbor, and hate thine enemy. But I say unto you, Love your enemies, bless them that curse you, do good to them that hate you, and pray for them that spitefully use you, and persecute you; that ye may be the children of your Father which is in heaven; he maketh his sun to rise on the evil and the good, and sendeth rain on the just and the unjust. For if ye love which love you, what reward have ye? Do not even the publicans do the same?

Matthew: Chapter 5, Verses 43–46

With Buddha:
Let a man overcome anger by love, let him overcome evil by good; let him overcome the greedy by liberality, the liar by truth! If a man foolishly does me wrong I will return to him the protection of my ungrudging love; the more evil comes from him, the more good comes from me; the fragrance of goodness always comes to me; and the harmful air of evil goes to him.

Instead Of Violence, pg. 471

With Lao-Tsu:
Wherever you advise a ruler in the way of Tao, Counsel him not to use force to conquer the universe. For this would only cause resistance. Thorn bushes spring up wherever an Army has passed. Lean years follow in the wake of a great war. Force is followed by a loss of strength.

Verse Number Thirty, Tao Te Ching

Most of all these meditations teach me of forgiveness, of all human qualities, the hardest to achieve, perhaps the closest to the divine?

To turn an enemy into a friend.

Deny our capacity for animal rage.

Isn't this what Bela Tiffany and William Green have done?

I watch Bela Tiffany and William Green, men of my blood who once tried to kill each other with the weapons of their day. I know their memories, the same as mine. They saw great armies move in the night. Shouts. Screams. Rebel yells. Guttural curses. Artillery shells whisper above their heads, a terrible thunder. Distant figures fire from smoke-hidden hills. Rifle thrown to the shoulder. Lead the running silhouette. Aim at the belly. Caress the trigger. Fire. Stock slams against the shoulder. Figure sprawls. Load. Fire. My GOD! A bayonet gleams out of the smoke. Parry. Thrust. Attack. Bayonet extended. Arm jerks. Sweat in the eyes. No blood. Smash the butt. A jaw? Who screams? Do I? Another thud. A screech…and the breath sucks into aching lungs. Did we win or lose? WHO CARES? WHO THE FUCK CARES? I AM ALIVE!

FORGIVENESS…

And I watch the two stare at each other.

And I know their minds.

For all my years I have longed to find the German gunner who fired the 88mm shell that blew my life away.

Find him.

Reach out.

Shake his hand.

That's all.

Shake his hand.

The war is over.

Done.

Enemies no more.

And I see their hands extend, Bela's and Bill Green's, a short, sharp movement of their shoulders, palm to palm, flesh to flesh.

I hear no words.

None required.

Forgiveness.

The lesson taught in this monastery.

That humans can live in loving groups.

Is that not the most important lesson I can learn?

Convert my enemies into friends.

The world spends so much time in enemy-making.

A moment each day spent in forgiveness might calm our anger and our rage, our fear of enemies, our need for weapons.

A moment plunged deep into silence, a meditative peace, might lead us to such compassion.

15 September 1984
The Cathedral
Chartres, France

I've left Paris for Chartres. I still live with this terrible conundrum, formed by my travels through these fields of France where I fought as a boy in 1944: how does a man live without a gun in a world of enemies and killing machines? Can there be no peace? Must we kill and kill and kill? Think of all the millions of people killed in war in my lifetime. Think of all the animals I have eaten in order to live: sheep and cows and trout and crabs and lobsters. The birds I shot. Am I but a mouth and anus and prick to ejaculate sperm, all the rest of my pretensions a joke?

I've settled in this town near the most beautiful building I have ever seen in my life. I plan on finishing my first draft of *On Being Wounded* here. This Cathedral surpasses others I have seen: Reims, Notre Dame, Salisbury in England, Santiago de Compostela and León in Spain. Its vast arched open spaces in the vault rise toward a sense of mystery and majesty as if the ceiling were a heaven and the stained glass windows were the moon, the sun, and the stars sending shafts of blue and red and gold and yellow into the air—planets alive in the misting light.

I live at my bed and breakfast, come here to the Cathedral each morning, meditate, then go back to my room to write, sometimes an afternoon train to Paris for dinner.

The miracle is that I have come here to Chartres, this Cathedral of flying buttresses, stained glass and light, a mystical place, dedicated to the mother, Mary, her compassion and her caring at the core of this Cathedral.

These cathedrals of the Middle Ages speak of other times, eras far more violent than our own, but times when places devoted to compassion were constructed over generations, as long as three hundred years. Kings and Lords gave money for the building; burghers and members of guilds donated stained glass windows; peasants gave labor to cut and haul the stone.

Whole societies built these artifacts of beauty, places of loveliness, dedicated to compassion and to grace.

Compassion linked with beauty in ways I do not yet understand, foreign to the American National Character.

2 and 3 April 1993
Pensione La Scaletta
Florence, Italy

This is the second time I've settled in Florence on this trip. I lived here for a month in February, then Rome, Pompeii, Naples, Athens, Crete, Delos in the Greek Islands, now back here again. The human created beauty of every place I've visited stuns me, Florence most of all. The whole city is a work of art. I cannot turn a corner, enter a plaza, look up into the air without stumbling upon a sudden vision of beauty.

I drink it in.

Yet, when all this loveliness was created in the fifteenth and sixteenth centuries Florence was one of the most violent places in the western world. City States fought each other with murderous intensity. Families lived in towers. From these sanctuaries they burst forth, determined to destroy their neighbors, enemies who lived next door.

Where did they find the energy in the midst of such violence to make so many artifacts of beauty?

Why is their creation so different from ours? Our artifacts, our American cities, lie ugly upon the earth, downtowns cold-gleaming high rise towers, ghettos cesspools of poison and violence, and the suburbs stamped from one mold.

Beauty forms each place I've visited, beauty formed from violence. Greece in the age of Pericles one of the most war-like states that ever existed, yet its buildings, statues, vases...and the plays, oh, yes, the plays: *Aeschylus* and *Euripides* and *Sophocles* and *Aristophanes*, tragedy and comedy invented by these men, Sappho, the feminine poet, whispering in her pure voice of love.

From their violence they made beautiful things just as the citizens of Chartres built the Cathedral in a violent time, as Goya, who lived in murderous eras in Spain, showed those times to us, as the Anasazi in Tsankawi where I lived in New Mexico, broke the pattern of their artifacts so that God and beauty could enter.

Some seek compassion and grace out of suffering. Others convert their pain to beauty...together they form our greatest human characteristic.

10 April 1993
Cathedral
San Miniato al Monte
Florence, Italy

I come here to meditate in the crypt every morning. The silence has an even greater resonance than Chartres, formed from time, I think. This a tenth-century crypt: for 1,000 years worshipers have prayed and knelt within it. One day I even stumbled into its deepest recesses by accident. A back door opened: I wandered into a great red bricked room, roofed with arching vaults. Its stillness, its dimness reminded me of my cave in Tsankawi.

I suck up the silence here, wishing I could store it within: the time has come for me to return to the United States.

I take home from these towns and places of Europe a sense of beauty, an understanding that beauty may be all that lasts. Our efforts at compassion, our loves, our hates, our acts of violence, these disappear with each generation to be repeated by the next, though we scarcely even understand how our actions impact our children.

Artifacts of beauty become, perhaps, humans' only eternity.

Whatever we mean by the word God—and I grow less certain of its meaning as I age—these paintings, sculptures, buildings let the spirit in, the patterns of our greed and rage broken when the miracle of beauty enters.

Beauty links with compassion then, ways to face the violence of our time.

Both rest in nature.

All three link in meditation.

These, then, the miracle of my life, my path beyond the weapon of my fathers.

Out of these voices of my ancestors, my past, those Chronologies, Rifles of My Fathers, People of Peace, the writing of Levi D. Tiffany, my Journals have I been able to throw away my weapons, construct this new ethic by which I now attempt to live.

Most of all it is an ethic without enemies, filled with compassion and forgiveness.

It is an ethic where, truly, the meaning of *E Pluribus Unum*—one out of many—is finally known.

It is an ethic of equal opportunity where the rights of inherited privilege are restrained.

It is an ethic where, finally, the bias of race and gender has been abolished.

It is an ethic where the wonder of the natural world is protected and where I live in it as much as I can.

It is an ethic of beauty, both natural and human made.

And, so, it is the simplest of all ethics, an expression of the purest principles of my People of Peace.

It is the tradition of another America, one we must heed if we are to survive.

The Fourth Chorus

William Bigbee Green

William Green, now in his old butternut uniform, rejoins Lydia, Levi, and Bela, their voices a soft murmur, his words sharply punctuating their gentle and indistinguishable chorus.

The thing is, boy, you are on the right track, only you still ain't far enough down the road yet.

You got to figure out why men commit such violence over over and over again.

I didn't understand nawthing until I was over seventy-five.

Nawthing.

But, then I begun to see her.

See her straight and clear.

You on the right track, that's for sure, to figure out what to do about the violence and the wars.

Enemies.

Weapons.

That's what it's all about.

And one more thing, boy, one more thing you ain't figured out yet:

SHAME.

Yup.

Shame.

That's what drives us all. Every man who's killed for whatever reason. Glory. Patriotism. Comradeship. Hate. Fear. Jealousy.

Shame.

You don't think so?

Well, lemme tell you.

I got disattached from my Alabama Command in early '64. Jus' plain lost. A damn fool. Since I had a horse and was in the cavalry I joined Bedford Forrest's troops. Now, Forrest was what you might call a mean man.

Just a plain mean man.

I was with him at Fort Pillow.

I saw what he did.

He let his troops go.

Let 'em kill all those blacks in cold blood.

Cold blood?

Hell, boy, there ain't no such thing as cold blood. It's hot. It sprurts out. You oughta know the way you was shot, all that blood spurtin' from your head. Only nineteen weren't you? Well, I was twenty when they shot me at Munfordville. Blood poured from me like the great pools of blood that spread from those three hundred colored soldiers we butchered at Fort Pillow, the fifty or so whites.

Murdered.

The same way we lynched blacks in reconstruction and after.

Why do you think I did that, boy?

Rode with the Klan for two long years until, one day, I just up and quit. No matter how they hasseled me, I never rode again.

Ya know what, boy?

Ya know what?

[His old body hardens. He seems to shake.]

Ya know what?

We were all shamed.

It was our shame drove us to do more.

Kill more.

Rape women.

Hurt little children.

That's what drives men in war to do the worse things: shame.

Ashamed of who they are and what they have become.

Ashamed they have taken the life of another human being.

Ashamed down to their roots.

The words, patriotism, honor, glory…all those big words hide their shame. They're killing another human being. Breaking Christ's sweet law.

They got to be ashamed.

But admit it, boy?

Admit it, they'd have to stop.

And they can't stop because to admit it would destroy their image of who they are.

So they make up big words to hide behind. God. Country. Comradeship. They find new enemies. Invent new weapons. Kill more and more and more.

I know, boy.

I did it.

Did it for fifty years.

Blamed the damned Yankees. Made 'em out as enemies. Not human. Hated black folks. Beasts. Had to be kilt.

But, now, I see.

They were just people, people like me, Bela over there, doing what he believed was right. Turn a man into an enemy, you change yourself into a demon, do awful things.

And, now I try and forgive myself at the end of life.

What we did in the south was bad.

What the north did was bad too…but it had some humanity on its side. It wanted to free the blacks we held in slavery or we kilt.

It's all bad, boy, it's all so bad.

You're on the right track. Forgive enemies. Throw away the gun.

For you were ashamed too. A feared you were a coward 'cause you got shot after one day. But, hell, boy, you weren't no coward. You did what you were supposed to do. But it took years to get over the shame, didn't it?

Every man who ever fought is ashamed.

Tell of shame to others.

Teach of enemies and weapons.

Enemies. Weapons. Shame. Give them up for redemption.

You do it, boy.

You do it.

Maybe you can be the first.

And, from the shadows, Levi's, Bela's, and Lydia's voices suddenly swell up, chorus with his.

It's your job, son.

It's your task.

Deal with the violence. Live beyond it. Live by those words you so treasure, by the beauty. Search for compassion and tenderness.

Your job.

The task your ancestors will to you: me from the south with that past of killin', Levi here from the north with his refusal to kill, Bela, his brother, who killed for his just cause and hated it, me who killed for a cause proved false. You got all these in you, boy. Together, they lead to a real understanding of America and maybe, jus' maybe, the hope for change.

CHAPTER SEVEN

And What Are We to Do?

And what is it I have learned from this immersion in my past as an American for these many years? What is it I have learned from my ancestors, their enemies, their wars and weapons, and from my People of Peace?

That from those men who gave me my life I received great lessons of courage: Solomon Lombard breaking from his past to take part in a rebellion that formed the United States; William Green and Bela Tiffany, wounded in the war that made this country into a Union; Levi Tiffany, saying "no" to our imperialism; my father, daring to fly military airplanes before there were parachutes.

The courage of men who risked their lives for their nation.

In some way I do not understand their courage was passed to me. Never mentioned. Never taught. But from childhood I understood: a man's role was to die for his country if the moment of history arrived. Their courage must have been mine as I stood on the banks of the Canal des Mines in France, groping my way down that destroyed bridge under fire.

I learned from them the spirit of the wilderness: Levi Tiffany and William Green, both farmers, long before there were machines, a love of nature, wilderness to them part of the Divine.

From Levi and from my People of Peace compassion, caring for the weak, the neglected, the poor.

From my love of wilderness and from my People of Peace, the writers I read, the painters I observe, my love of beauty, words my choice as a medium, building blocks for books and novels, poetry and fables.

From all these ancestors, from all these People of Peace, the courage to give up the weapons of my fathers, form a new life beyond those weapons.

Ancestors, People of Peace, gave me the understanding that, to give up violence, I must go far beyond my refusal to own a gun, recognize that it is the world where enemies are made I must avoid, create my own ethic for living in a world of violence.

I must learn to care for nature and for people in small ways, not the arrogance of great movements, but in the love of a woman, caring for my children, perform acts of kindness to both strangers and my friends, discipline myself to show good manners, even when attacked and riled by unknown strangers on mean streets in quick-darting automobiles, will myself to acts of forgiveness for those who injure me, widening the circle to turn them into friends.

Attempt to make beauty in small ways in my novels, poems, fables, in my volunteer work at the Denver Art Museum, helping children and adults see the loveliness, far from violence, in sculpture and painting.

Grow beyond my shame.

The shame of combat we all have known, which William Green so poignantly pointed out to me. The shame of killing when we bomb small countries without risk to ourselves, without casualties.

Force ourselves to recognize that great dilemma, the one posed by my great uncle, Bela Tiffany, and by Justice Oliver Wendell Holmes: "society has rested on the death of men." Understand that such deaths occur from many different motives: when the innocent are threatened as in slavery or in the Holocaust by real enemies, those of evil intent; when self-interest predominates as in our centuries-long wars with the Native American or the war with Mexico in 1846; when self-defense is required, from the attack of the Wampanoags in 1675 to the terrorist strike on September 11th.

Recognize and accept that in all three instances, some will object, turn to pacifism.

Understand that sometimes in some wars, all three motives mix in ways that cannot be separated. What was the Spanish-American War about? On one level, self-defense after the battleship, *The Maine,* was blown up. On another, protection of innocents, those people cruelly dominated by Spain. On another, extension of the American empire.

Given the nature of our modern weapons, the ease with which we kill at a distance, the proportionality of our response to all forms of violence and kinds of wars has become the most complex moral problem of our era. When attacked and we need to defend ourselves, is our response to an attack proportionate to its force?

When the Wampanoag Indians attacked in 1675 we completely destroyed them and their allies with pike, Snaphaunce, fire, and slavery. In the Civil War, we lived by "unconditional surrender," demolishing the land, the crops, the infrastructure of the south. In World War Two, though we were so clearly in the right, we retaliated with terror bombing, the killing of innocents, techniques we continued in the war with Vietnam.

Our tradition of over three hundred years is one of overreacting, total destruction of our enemy.

We do not even learn from current conflicts where we see enemies destroy each other. We watch Israel and Palestine, where acts of terrorism lead to acts of revenge which bring about more acts of terrorism, then greater blows of revenge...cycles repeated senselessly as we support and arm one side against the other, refusing to insist that each side has turned the other into an enemy of "total evil," a scapegoat, refusing to seriously play the role of peace-keeper, accepting the risks and the sacrifices required.

The enormous danger is that, when attacked by ruthless enemies as in the case of the World Trade Center, our anger and paranoia take control: we turn merciless with acts of revenge, raising the struggle to another level. The "enemy" comes to symbolize all evil and must be obliterated so that the world will, finally, be quit of its evil. We have created a scapegoat we must brutally destroy. In thrall to "unconditional surrender" and "no duty to retreat" we crush our enemy in fits of rage, using the most powerful weapons we possess.

We do not know how to think about enemies in this country. Because we have so much power, we use it heedlessly, plunge into this ceaseless bombing of smaller countries, without risk to ourselves, never considering the most important reality of any war, "unintended consequences." We rush to the invention and production of new weapons, the madness of missile shields. We never look at the hidden causes of our rage: our fear of enemies, our shame over what we have become.

We have never learned to think about enemies in America, how to recognize the real enemy, how to understand when we sink into illusion and fantasy, terrified of enemies who do not exist. We did that in the Civil War when Congress sacked generals, accused them of treachery. We did that in the great witch hunt of 1919 when Attorney General Palmer rounded up the "reds" and sent them to Russia. We did that when we put Japanese civilians in concentration camps in World War Two. We did that in the days of the McCarthy persecution. We did that in the Cold War when we created 30,000 weapons, enough power to destroy the world again and again and again. We did that in Vietnam, Central and South America. We overwhelm Afghanistan, killing innocents with our weapons of mass destruction. In deference to fear of our enemies, seeking security and safety, we threaten our civil liberties.

As a nation we have perfected out of our many wars our weapons that kill at a distance, defeating our enemies, destroying the earth, its infrastructure, its people without costs to ourselves, triumphing in "unconditional Surrender." As individuals we follow this precedent, use hand-held weapons of overwhelming force, spray imaginary enemies with round after round, blow innocents up with hidden explosives, wedded to "no duty to retreat."

With little sense of history, each generation is certain the enemies of its day are unique and must be destroyed.

Each generation must invent new weapons, its "own" weapons to defeat those enemies.

The making of enemies, using ever-increasing fire power to destroy them, lies at the heart of much of our violence. We have a capacity to sometimes create enemies who do not exist or inflate real enemies into demons with magical power. These "scapegoats," never make massive threats to our being but are the enemies endlessly rediscovered in cycles over our long history, to be destroyed each generation by ever more sophisticated weapons.

Our most important tasks, if we are to control our growth of weapons and halt our repeated cycles of terror, domestic and foreign, are, *first*, and highest priority, to develop our own individual ethic for a life beyond violence, enemies, and weapons; *second*, to root our life in this other tradition of America, its People of Peace, a tradition unknown, neglected by our mainstream; *third,* learn to recognize which enemies we and our nation make are real, which false, formed by our paranoia or self-interest, then, how to react proportionately to threat and attack; *fourth,* to reduce our dependency on weapons, finally seek to give them up; and, *fifth,* to deal with our shame.

These five tasks, then, lead to a life beyond the weapons of our fathers.

Searching for an Ethic

The burden on each of us is to examine our lives, see where they mesh with enemies and violence, create our own ethic to live beyond these conditions.

For some of us this will be hard.

Our jobs, our careers, even our family lives are sometimes wracked with violence, encircled by enemies we fear, whether real or not.

We see psychotherapists, take pills, toss on the bed at night, pray, walk the floor, yet nothing seems to change. Our lives continue, spinning down into dark funnels with increasingly narrow rims.

For me it took the disaster of a broken marriage, a collapsing career to finally recognize that the burden was myself. No new woman, no new career, no new family could change the self that lay within. I had to change myself, not through psychotherapy, through pills, through new relationships but by plunging deep within, facing what was there, forming a new self out of my past, an act finally of moral will: I simply had to change or be a drunk, die of a heart attack or commit suicide.

I gave up my belief that the problem of the violence in which I lived, within and without, the image of the enemies that haunted me, lay outside myself.

It is so difficult to admit that our violence, our need for enemies, our rage to destroy, our ability to revel in the process of enemy-making rises from our own human condition, from deep within, from ourselves, how we are raised, treated by our parents as children, our birth position, the stresses of our culture, the traditions of our past. We want so much to have evidence that violence and war are not part of the human heart. We want to hear that poverty is a cause, injustice is a cause, government is a cause, our parents a cause, families, communities, schools a cause. If we can discover a cause outside ourselves, then we can reform that condition, change reality and, behold! our violence vanishes, a new utopia will be formed.

We do not deal with our own part in making enemies.

We so long to believe that peace is possible, a condition beyond our violence, if only we can enforce enough "shoulds."

I do not know if my new self created my ethic or if my ethic created my new self.

In a time of intense inner turmoil and loneliness I changed, began to write, sought beauty, lived in the natural world, formed an ethic out of my past, my *Journals*, from those People of Peace discovered, out of the wisdom and the pacifism of Levi Tiffany, the lives and voices of Bela Tiffany and William Bigbee Green.

From that revolution in my life has come a rich fullness in my days and nights, an ethic beyond the weapons of my fathers. I published *On Being Wounded* after seven years of writing, am beginning to publish my poetry and have a novel for sale. I work as hard as I can at my relations with my three children: I like and admire them as well as loving them—they're great people! I have hopes that, someday, the pain of that failed marriage may yet be salved. I live with my companion of fifteen years, fulfilled in caring. I take tours of children and adults through the Denver Art Museum, seeking to bring them some sense of the wonders and the beauty of artifacts made by men and women. At seventy-seven I can still even trout fish or wade a stream.

All these rising from my acts of rebellion against my past, an ethic cut from flesh and soul.

To live beyond the weapons of my fathers, beyond violence, it is essential to root my life in my ethic: compassion, beauty, and the wonder of the natural world.

From that ethic comes the decision to give up our weapons.

If we decide to live without weapons yet lack an ethic to give us strength, that decision may not last.

Not that all ethics will have the same components. Mine rises from my own roots: my childhood in the out of doors, a Christian upbringing, my mother a painter, all these form its base.

For each the base of childhood will differ.

The task is to seek those ways rising from one's past, make a path for oneself from that beginning, out of one's deepest heart, beyond the weapons of our fathers.

That task is composed of so many parts: reading about issues of peace and war, violence and compassion; meditating on some of those readings; learning from one's past and from People of Peace; keeping a *Journal* on peace seeking; experimenting with personal acts of peace.

Each person will organize these several aspects of creating an ethic beyond the weapons of our fathers in different ways. I've listed in the bibliography following some of the books that have helped me in my discovery of my new path. I hope they may prove fruitful for others.

Honor Our Tradition of People of Peace

We need to honor—more than simply recognize—the tradition of our People of Peace.

It took me years to accept their importance to America.

In early drafts of this book I tended to denigrate those who devoted their lives to march to Thoreau's "different drummer." Recurring cycles of violence convinced me that those who longed for a better world would always fail.

The War in Vietnam, rebellions and *coup d'etats* in Central and South America, murderous rages in our own central cities, continuing bias against African Americans, Native Americans, Latinos, child abuse, domestic violence, serial killings, all seemed to me evidence of the cruel and dominant aspects of humanity.

No matter what actions were taken, cycles of violence always returned, the acts of each generation more virulent as weapons increased in obscenity. Yet slowly, so very, very slowly, as I read of People of Peace, their courage, as I learned I had an ancestor who said, "no" to the wars of his day, I began to understand: these people were the best the nation had given to the world.

After discovering Levi Tiffany's Testimony against the Mexican-American War, I stumbled on a book, *Power to the People: Active Nonviolence in the United States*, edited by Robert Cooney and Hewlen Michalowski.

At first I rebelled against these "namby-pambies," jokes, cowards and finks.

But then I read of the military men: Ethan Allen Hitchcock opposing our treatment of the American Indian, his reluctance over the Mexican-American War, the officers who refused to participate in Chivington's massacre of Cheyenne at Sand Creek.

I thought of a man who once had been the greatest figure of integrity in my life: Brigadier General Jake Lindsay, who taught me truths about America in the 1950s that no one at that time even understood, much less said.

Once, after he left the Service, in 1952 or so, just as the nation was beginning its economic explosion, he looked at me, smiled, and said:

> *That thing America uses to measure itself, that ol' GNP, it don't mean nothin' does it, Eddie?*

It took me years to understand his meaning. The GNP only measures economic growth. A great country must be far more than its capital, goods, cash flow, interest rate and stock markets.

Yet in the world in which I lived money measured everything.

Not true of my People of Peace I discovered.

They lived by other Gods.

Yeah, [Levi Tiffany sighs] ... *lived by other Gods.*

That's what we tried to do, all of us before the Civil War, those you honored in your Roll Call, your voices of Peace, your Chronicles, the great ones of my day: Woolman and Penn, Thoreau and Emerson, Garrison and Douglass. Sojourner Truth and Lucretia Mott.

We wanted to form a Paradise on this new land of America.

Why, four states in the Revolution made conscientious objection a right in their early Constitutions.

Our hopes, our dreams for that Paradise died in the Civil War, all that killin', all those deaths.

But [his face broadened into a smile]...*it all come back, it all done come back.*

You got those same men and women in your time, people who care, who make a difference, people who say "no" to the violence and wars of their day, refuse the making of enemies.

Why, though they are not even honored in your time, scarcely known, they still are the men and women who struggle to create that Paradise in America.

Those are the ones you must speak of, the ones who lived beyond enemies and weapons, risked all to form that better world...the tradition of my youth in America, before the Civil War, has not been lost. It has resurfaced in your time in your People of Peace, their voices of Peace, their acts of Peace Making.

Tell us of them, those who still walk to Thoreau's "different drummer."

Tell of us of their acts, contained in your Chronicles.

Tell us of their voices of Peace.

Another Roll Call of a Few of My People of Peace

From the Civil War until World War One

Lawrie Tatum

Quaker, volunteered to be an Indian Agent at Fort Sill, Oklahoma, struggling to bring peace between warring Comanche and Kiowa tribes and the white settlers.

Ida B. Wells

At great personal risk, defied the accepted belief that lynching in the south came about because black men were rapists. She determined that blacks had been tortured and murdered for everything from "quarreling with whites" to "making threats." Moreover, children, as well as women, had been lynched. For her findings her newspaper office was looted and burnt and she was driven from the south.

William Jennings Bryan

Resigned as secretary of state rather than support President Woodrow Wilson's warlike policies before America entered World War One in 1917.

Eugene Debs

Labor and political leader. Though a Socialist candidate for president, he was jailed by President Wilson for his opposition to World War One.

Susan B. Anthony, Elizabeth Cady Stanton, Carrie Catt

And all those marvelous women who, for seventy long years, struggled for women's right to vote.

Chief Joseph

Chief of the Nez Perce nation, led his tribe on a 1,600-mile retreat from the U.S. Army. Finally trapped just south of the Canadian border, he surrendered. In Washington, D.C. he made a moving speech defending rights of American Indians.

Jane Addams

Founder of the Settlement House Movement to help European immigrants achieve rights in America, also a leader in the fight for just wages, safe labor conditions, and the fair treatment of women and children. Helped form the Women's Peace Party, spoke out against World War One, and won the Nobel Peace Prize in 1931.

John Muir

Early conservationist, responsible almost single-handedly for the creation of Yosemite National Forest. Helped bring about formation of thirteen other national forests. Pioneer of the environmental movement.

Roger Baldwin

Conscientious objector to World War One and jailed for his opposition. Founded American Civil Liberties Union after the war.

Edward S. Curtis

Devoted a lifetime to photographing Native American nations just as they were overwhelmed and destroyed by expansion of western society.

And from World War One to Today

Eleanor Roosevelt

The wife of a president who, quite simply, made the nation a better place as an advocate for the poor, for people of color, for peace, and for human rights.

Jessie Daniel Ames

Saw lynching in the south as a great evil, founded the Association of Southern Women for the Prevention of Lynching.

Dorothy Day

A militant social activist, founded the Catholic Worker Movement, edited *The Catholic Worker* newspaper, struggled against the Vietnam War.

David Dellinger

A conscientious objector to war, beginning in World War Two. Jailed many times, he opposed the Vietnam War, and was an advocate of non-violence.

A.J. Muste

The grand old man of the peace movement, began his opposition to war in World War One. Forced to resign as a minister, he was a labor organizer and bitter opponent of the Vietnam War.

Maurice McCrackin

War tax resister, refused to pay that portion of his taxes that supported the defense budget. Finally jailed over his refusal to pay his war taxes.

Cesar Chavez

Founder of the National Farm Workers Association, to bring economic justice to migrant workers.

General Smedley Butler

Leader of Marines, participant in Boxer War in China, other forays into Central America, at retirement wrote his expose of the miltary, *War Is a Racket.*

Martin Luther King Jr.

And all those blacks who struggled, after a hundred years, to bring Civil Rights to African Americans and people in poverty.

Warrant Officer Hugh C. Thomson Jr.

Ordered his men to fire in order to stop the atrocities at My Lai in Vietnam.

Tom Rauch

Peace activist, who organized encirclement of Rocky Flats in Colorado where nuclear triggers for atomic bombs were made.

And all those many **conscientious objectors** to the Vietnam War, some jailed, some who fled to Canada, men who paid a huge price for living by their values.

And all those **Hispanics, women, gays, Native Americans** and **special needs people**, who have struggled to obtain equal rights, denied for so many centuries.

A Few of the Acts of Peace-Making and Compassion Committed by These People of Peace From the Civil War to the End of the Twentieth Century

1869–1919	Founding of Suffrage Associations, fifty-year struggle for women's suffrage.
1877–1894	Years of labor strife as nascent labor unions struggle for economic justice against railroads, iron and steel companies. Soldiers used as strike breakers.
1898–1904	Thomas Reed resigns as Speaker of the House in opposition to Spanish-American War. Opposition particularly strong in New England.
1915–On	Major peace organizations, still active today, formed: Women's International League for Peace and Freedom, Fellowship of Reconciliation, American Friends Service Committee, American Civil Liberties Union.
1917–1918	4,000 men refuse to serve in World War One.
1918–1920	President Wilson attempts to have United States join League of Nations.
1931–1936	Great surge of pacifism.
1940–1945	52,000 men refuse to serve in the military in World War Two.
1945–On	Resistance to military use of atomic bomb begins immediately after the war and carries forward for fifty years in all kinds of ways: picketing, civil disobedience, marches, refusal to pay taxes.
1947–On	CORE initiates first Freedom Ride through the south. Twenty-five years of struggle for Civil Rights begin, culminating in passage of Civil Rights legislation.
1960s–On	Both Hispanics and Native Americans begin to organize to demand their Civil Rights.
1962	Rachel Carson writes *Silent Spring*.
1964–On	Protests against Vietnam War begin, escalate into major confrontations. Peace movement bitterly divides into violent/non-violent factions.
1966	Women's movement begins struggle for equal opportunities in work, career and pay.
1970s–On	Protests against nuclear weapons increase in intensity, culminate in New York march in 1982. Environmental Movement begins in earnest. Struggle for Gay Rights begins and intensifies.
1980s–On	Based on our support for right-wing governments in Central America, opposition culminates in the Sanctuary Movement.
1980s–On	Exploration of a whole variety of approaches for reaching peace occur: "win-win concept;" conflict management, alternatives to violence taught in schools and prisons.
1980s–On	United Nations experiments with a peace army.
1990s–On	Environmental movement continues its struggle against destruction of air, water, and land resources. Gay Rights consolidates its gains.
1994	Brady Bill passes, slowing illegal purchase of handguns.

Most important, throughout this half century a new coalition is slowly formed, composed of groups demanding protection of the environment, groups seeking human rights, and groups seeking economic equity. This coalition comes together in Seattle, Washington, D.C., and Genoa, Italy opposing the World Trade Organization, G-8 Nations, and global corporate power.

Voices of Men and Women Who, as Peace-Makers and Lovers of America, Continue the Vision of the Land

If the white man wants to live in peace with the Indian he can live in peace. Treat all men alike. Give them the same laws. Give them all an even chance to live and grow. All men were made by the same great Spirit Chief, they are all brothers. The earth is the mother of all people, and all people should have equal rights upon it....

Chief Joseph

The slow progress towards juster international relations may be traced to...Grotius...Kant...Tolstoy....Each in his own time, because he placed law above force, was called a dreamer and a coward...but each did his utmost to express clearly the truth that was in him....

Jane Addams

I helped make Mexico safe for American oil interests in 1914. I helped make Haiti and Cuba a decent place for the National City bank boys to collect revenue in. I helped purify Nicaragua for the international banking house of Brown Brothers....I brought light to the Dominican Republic for American sugar interests in 1916. I helped make Honduras "right" for American fruit companies in 1903. Looking back on it, I might have given Al Capone a few hints.

Major General Smedley Butler, USMC

...the single most destructive aspect of existing society [is] the willingness to pursue one's own fulfillment without concern for the fulfillment of one's fellows....

David Dellinger

There is a certain indolence in us, a wish not to be disturbed, which tempts us to think that when things are quiet, all is well...human beings acquiesce too easily in evil conditions; they rebel far too little and too seldom....

A. J. Muste

There is no peace because there are no peacemakers. There are no makers of peace because the making of peace is at least as costly as the making of war—at least as exigent, at least as disruptive, at least as likely to bring death and disgrace in its wake.

Daniel Berrigan

They want America to trust the world and are sure that the world will in turn trust America. Their faith is too naïve. They do not realize that a nation cannot afford to trust anyone if it is not willing to go the length of sharing its advantages. Love which expresses itself in trust without expressing itself in sharing is futile....

Reinhold Niebuhr

The policy of the government...during and after the war, say in the bombing of Hiroshima and Nagasaki, made it clear that to provide scientific information is not necessarily an innocent act. I do not expect to publish any future work of mine which may do damage in the hands of irresponsible militarists....

Norbert Weiner

The peace movement knows there is something fundamentally evil about this society....All the years of killing in Vietnam. All the murderous weapons being sold throughout the world. All the endured violence of Civil Rights struggles and the freedom rides and sit ins. Through all this one comes to know the seriousness of the situation and to realize it's not going to be changed just by demonstrations. It's a question of risking one's life. It's a question of living one's life in drastically different ways.

Dorothy Day

Levi Tiffany, with great satisfaction

There. That lady, Dorothy Day, just said it right. Living one's life in drastically different ways. That's the only way true change will ever come about.

And yet…

This tradition of People of Peace of which Dorothy Day and Levi Tiffany speak is largely ignored as a tradition in the United States.

We hear little of it in this moment of a new war.

90% of the people are for violence to capture or kill bin Laden and the al Qaeda, destroy the Taliban, control terrorists. Opposition to the war is barely noted in the mainstream media, often only expressed on the Internet.

Those with reservations about the war are viewed as unpatriotic: is a new McCarthyism eminent? What about the threat to our civil liberties posed by the actions of the administration?

It is difficult to sense at this time how the People of Peace in America of this day will coalesce into some meaningful force, how they will express themselves in this time of terrorism directed at our own nation.

But I remain certain that new ways of speaking will arise: the great traditions of over three hundred years simply will not be lost. New voices will be discovered, speaking to the conditions of the day. They will wrestle with the preservation of our civil liberties in times of terrorism. They will deal with the issue of the proportionality of our response to terrorism. Should we use such overwhelming force, without risk to ourselves? Does not our response, instead, require we take a long, hard look at ourselves, ask why we use so much of the world's resources while others starve, young men in Muslim countries so unemployed, young women treated so unfairly?

I doubt if the terrorists driving those planes ever heard of Martin Luther King Jr. or any of our People of Peace.

I doubt if Dylan or Eric in Columbine High School every heard of those American Army officers who turned in their commissions rather than kill Native Americans.

We desperately need to tell the world about this other America, not the one of consumption and military force but the one of fairness, equity, and the search for justice. We desperately need to build courses for high school students around such people and their contributions to the nation, not as weak and impotent but as people of enormous moral fiber, able to say "no," to the powerful and repressive forces of their day.

We need to make movies and TV programs to tell another story beyond these weary tales of violence and crudeness, instead of showing the world in the media the lowest common denominators of the human race.

Newspapers need to return to the fact that there are and have been such people in America: my guess is that they are never mentioned in the schools of journalism.

If we really want to end our violence and the world's, why not trumpet our tradition of the People of Peace, of compassion, of caring, of widening the circle of democracy, of creating beauty out of violence, of preserving wilderness as a spiritual value?

Why not pay particular attention to the best of them, those who refuse to plunge into the world of enemies, who live by forgiveness, the aroma of love theirs, the taint of hate the aura of another?

Recognition and Understanding of the Role of "Enemies"

I have spoken so much of enemies in this book that it is unnecessary to repeat warnings of the danger of false enemies, formed from paranoia, or enemies made out of our self-interest.

To turn from them, to have the wisdom to discern real enemies, rises from the wisdom of that ethic created out of the very soma of our soul.

To forgive enemies, to broaden our circle of friendship, so they are included, demands an even greater effort of love.

To include enemies, broaden the circle of our democracy so that men and women of all ethnic groups, black, brown, white, of all sexual persuasions, of all ages, and all things of the natural world are brought within, out of the cold, represents the major task of our democracy.

The problem, the real problem, is to recognize the enemies for whom they really are, never create enemies where none exist in senseless causes for self-interest, for the sake of our paranoia, never act disproportionately as we have done again and again in cycles of our history: wars against the American Indian, Mexico, the Philippines, Cuba, Puerto Rico, Panama, Central and South America, the Caribbean, Vietnam, Grenada, Iraq, the "collateral damage" done to innocents in Serbia and Afghanistan, perhaps, in the future to others in the third world.

And when enemies are real, as they sometimes will be, we need to discover reactions more feasible, our application of force often inappropriate to the threat. Clearly we always need to act in self-defense, protecting our people, the Declaration of Independence, the Constitution, the Bill of Rights, the bulwarks of our democracy, from acts of destruction.

But we must refuse to listen to that terrible fear of the enemy so deep within that leads us to demonize those enemies, invent new weapons to destroy new enemies generation after generation.

We need to search for different ways beyond the weapons of our fathers, reconsidering those two ideas, which lead us to the use of overwhelming violence to destroy our enemies: unconditional surrender and no duty to retreat.

Reconsider the Role of Weapons in American Life

The macabre dance between enemies and weapons relentlessly continues.

Twenty-two years ago I gave up the weapons of my fathers and, since then, have struggled to live by my ethic: the ethic of the wilderness, compassion, beauty, failing sometimes, failing more often than not, failing miserably. William Green, Bela Tiffany, Levi Tiffany, Lydia Tiffany did not imagine a nation where the death of innocents from our bombs is so ignored, where our leaders speak carelessly of missile shields, not even knowing if they are technically possible, where any fool can own a gun that kills at the rate of hundreds of rounds a minute,

The Second Amendment to the Constitution was written to permit the possession of a single shot muzzle-loader, not weapons of mass destruction.

The slow and lurching movement toward the control of individual weapons must receive our continued and fierce support if we are to ever halt the senseless violence in which we live.

Just as much discipline must be exercised in the development of weapons of international destruction.

In our bombings, in our use of the atomic bomb in World War Two, we began to accept the idea that we, the United States, must always have greater technological capability to destroy than any other nation. We now carelessly accept the policy that we always must have more powerful weapons than any other nation in the world, *all of them our potential enemies.*

These new weapons used in the 1991 war against Iraq, in 1999 against Serbia, in 2002 against Afghanistan, our ability to kill at a distance, have created a great moral lacunae in our civil life. In these wars, we, American citizens, continued our daily rounds with little inconvenience: shopping, working, commuting, our usual chores, our usual leisure time, sports, TV, cocktails, making love. We, as a nation, have little knowledge of the pain our armada inflicts on others, our technological capacity to kill at a distance. We have come to believe we can kill with impunity, without paying a price.

Can we—*dare we*—lead the world from such weapons of mass destruction before the earth and its inhabitants are destroyed, perhaps the most important moral question of our time?

Yet this remains but a first step.

A far more difficult task awaits.

Each year new weapons for domestic use are invented, ever more obscene.

Each year the United States pours its excess weapons into undeveloped nations where killing remains an endless pursuit, often performed by boys of ten and thirteen.

Each year the United States spends enormous amounts of money on the research and development of new weapons of mass destruction, weapons to kill people, destroy infrastructure, obliterate the natural world, finally, today, weapons to kill weapons. Each year private manufacturers devote large sums to the development of new weapons for personal use, weapons capable of ever increasing fire power.

The common denominator of all these weapons, domestic and international, is the ability to kill at a distance so that the user is far from the direct experience of destruction and death.

If we are ever to control the glut of weapons on this earth, it is no longer enough to work toward simply controlling their use.

Simply put, we must turn as well to the control—the halting!—of the development of new weapons.

Each generation wants to discover its own technologies of killing, its own inventions to destroy the enemies of its day.

Once we feared the USSR as the enemy. Now it is North Korea, Iran, Iraq, domestic terrorists.

New enemies to be destroyed each generation by new and more virulent weapons.

Dealing with Our Shame

And yet I do believe that we will never be able to give up our enemies and control the production of weapons until we face our shame.

I now am certain that all men who ever experience serious combat commit some act for which they are forever ashamed. It may be like myself: totally alone, thrown into the pitiless place of combat with little training and no knowledge of how to act, wounded with no chance of self-defense, naked to my enemies. It may be acts of commission: simply the act of killing another human being, the worst thing in the world a man can do, or even harder to accept, killing prisoners, sometimes a self-inflicted wound. It may be acts of omission, skulking, avoiding the front, failing to help a friend, running under fire, even refusing to fire.

It makes no difference. Combat creates conditions that lead to acts of inhumanity, cowardice, failing ones' companions, acts that shame forever.

Shame plays an important part in psychiatric failures in war. The combat soldier sees things no man was meant to see: bodies dismembered, friends killed, emotions shattered. He becomes ashamed of what the human being can do to his fellow human beings, what he has done, seen done.

Yet that shame cannot be admitted, hidden under the rubric of enemies and weapons, of heroism, of scapegoats, of conquest, of medals.

Shame haunted me for years after I am came home from France, the shame of the wounded. Only in combat for a day and a half, I was scarred by the fear that I had been a coward. What had I done in the war? Lugged some machine gun ammunition, fired one shot in anger, sat in a half-track, walked down a gravel road at night, fired on by Germans? My head had been blown open, my back and buttock shredded, shrapnel left in my hip: I was shamed to my very being by my failure, my wound.

My shame was so enormous that I threw away my medals, even my Purple Heart: how did I deserve a medal for being wounded the way I was?

Forty years of an unconscious shame passed before I returned to France, saw the place of my wounding, realized that I had actually performed the duty, performed it under the most trying of circumstances, without a friend, without one kind word, left alienated and alone to learn the way of combat myself, my acts were actually ones of greatest courage, simply doing what I was told to do, almost a super-human task. Only when I read of the Army's replacement policy, how there had been thousands of us thrown into combat the same way, without proper training, many killed within an hour of reaching the front, many captured, did I begin to understand that I had simply been the victim of a bad policy.

My dialogues with my forebears helped so much; I sensed their tolerance and their love, their understanding: had not they faced the beast themselves, struggled with their own shame?

Once I faced my shame, I could wear my medals with the greatest pride.

Yet no matter how hard one tries, a residual of shame always remains after acts of killing and wounding begin. Even after all my efforts, my dialogues with my forebears, my readings, my efforts at constructing an ethic beyond enemies and weapons, my search for the tradition of People of Peace, even after all these efforts and all these years, a residual of shame remains, always there, oh, no longer that fear of cowardice: that is gone. I know what I did. But the shame is simply pressed into my flesh by my wound, those great scars in back and buttock, that hole in my head, that piece of shrapnel in my hip.

The shame of the scar will never disappear.

The shame of acts of war.

William Bigbee Green and Bela Bentley Tiffany knew that.

They helped me learn it.

Enemies, weapons, shame, the trinity that leads us down the pathway to hell, our efforts at creating a personal ethic to turn from war and violence, our tradition of People of Peace, the two ways we humans have forged to move beyond that trinity, such fragile tools in face of such powerful forces of inhumanity.

Yet all we have; our honor and our hope.

Yet tools so severely neglected in this world we have made where enemies, weapons, and shame—sometimes the lack of it!—still seem to control our ways of life.

We now bomb smaller nations with impunity, killing innocents at a distance.

Kill them while we shop and go to fine restaurants, eat fancy dinners, look at TV, make love. Kill them without any sense of shame.

And, because we have no shame over the ease with which we kill at a distance, no understanding of the pain our bombs inflict, we continue killing and killing and killing, unable to recognize our own inhumanity, unable to search for more proportionate reactions to attack, unable to seek a lasting peace, only able to retaliate with massive weapons, gleeful at our power.

They are the inhuman enemy, aren't they? Deserve what they get? In the oddest kind of way we end by blaming the people we kill for their death: we are free of fault.

Perhaps the most important act we might undertake is, finally, to admit our shame: over all the innocent Germans and Japanese bombed in World War Two; for all the innocent Koreans killed in that war; for the Vietnamese and Cambodians bombed from 1965 to 1973; for the innocents maimed and destroyed in Iraq and Afghanistan.

We will never grow beyond the weapons of our fathers until we admit our shame, turn from making enemies, incessantly developing weapons to destroy them, until with Frederich Nietzsche, we finally say:

> *And perhaps the great day will come when a people, distinguished by wars and victories and by the highest development of a military order and intelligence, and accustomed to make the heaviest sacrifice for these things, will exclaim of its own free will, "We break the sword," and will smash its military establishment down to its lowest foundations. Rendering oneself unarmed when one has been best armed out of a height of feeling—that is the means to real peace, which always must rest upon peace of mind; whereas the so-called armed peace, as it now exists in all countries, is the absence of peace of mind. One trusts neither oneself nor one's neighbor, and half from hatred, half from fear, does not lay down arms. Rather perish than hate and fear and twice rather perish than make oneself hated and feared—that must someday become the highest maxim for every single commonwealth too.*
>
> ***The Wanderer and His Shadow***

How simple this all seems.

How simple it all is.

It is not the subject for academic research, for long papers, endlessly dry.

It is not the self-inflated flatulence of our politicians, their words drowning the truths of the courageous acts of the past.

It is an act of moral will.

If we had the moral will, tomorrow, we could enter upon a new path to halt both domestic and international violence.

With my ancestors and our People of Peace perhaps we can yet discover that will.

* * *

I ask myself if my search and the paths I have followed will ever be of any use to other human beings, bring about any change in our stubborn condition.

I am unsure.

I can only be certain that this search and its reporting have been as honest as I can make them.

And that is all of which I am sure.

I hope its honesty will appeal to others on their own journey.

Will the actions I have taken change anything?

What arrogance to even think of the question.

My search for a life beyond the weapons of my fathers, the lives and traditions of my People of Peace, my readings, all have taught me that one must not act from the expectation of results but, rather, a sense that the action is right and proper. It is in that act of moral will that the truth is finally discovered. The hope for change from one's effort is a false God, leading in directions of compromise, fantasy, and, finally, cynicism.

We discover the capacity to really live when we have given up the hope that the world will change from our actions, when we act only because we know, "it is the right thing to do." We become arrogant when we expect rewards for our good efforts. We become too anxious when we must have a victory. We do not write so well when it is for publication and not from our deepest soul.

For so many years I believed with George Fox, the Quaker mystic, that…"an infinite ocean of light (flows) over the sea of darkness."

I no longer am sure of that. I have seen too many lights put out in my years.

Yet I still nourish a faith in light. But I see it as dim, wavering, sometimes almost totally extinguished by the great sea of darkness, only kept aglow by the actions of men and women like my People of Peace, by the efforts of artists to make things of beauty, by the gleam of the sun on a summer morn, the song of the wind on my flesh, by the silence of my soul in stillness.

All I can do in this era of anxiety to diminish my despair, a citizen of this land of "enemy making" with its long nights of darkness, is seek to live by that wavering light within the natural world and its beauty, man-made artifacts of beauty, struggling to make a thing of beauty myself, break the pattern of my day so that I live in simplicity outside consumer America, discipline myself so as to control my anxiety and my ability to make enemies, live outside shame, perform unexpected and unannounced acts of compassion, never expect to find hope in human institutions, attempt in my own way to increase the light of beauty and compassion in the world, seek to understand the causes of our violence and our need for enemies, share whatever I discover, try and reduce the number of weapons in America.

We must live with our honor, not by our hope.

The Final Chorus

The Reunion of the Soldiers of One Family Who Fought in America's Wars as Citizen Soldiers

King Philip's War—Humphrey Tiffany
French and Indian War—Consider Tiffany
The Revolution—Solomon Lombard/Peter Zachary
The Southwest Frontier—James Lombard/Stephen Lombard
The Civil War—Bela Bentley Tiffany/William Bigbee Green
World War One—Edward W. Wood
World War Two—Edward W. Wood Jr.

And an Interloper—Levi Darwin Tiffany

Is this a dream?

A fantasy?

Or like Dante's *Inferno?*

I do not know.

But, now, we are all together for a moment, all men who share one pool of genes, all who have fought for the nation, citizen soldiers in ALL those wars, citizen soldiers joined together at a last encampment on a great plain, I know not where. The smoke of battle drifts over us. In the distance we hear the cruel mutter of shell fire, the occasional rap of small arms.

We lie tense but at ease.

Humphrey holds his Snaphaunce and his pike, Consider and Solomon carry their "Brown Bess," Peter Zachary, his Kentucky Rifle, the Lombards, their Plains Rifle and Colt Repeater, Bela Tiffany and William Green, their .58 caliber rifle, my father has strapped on his .45 caliber pistol, I carry my M–1 Garand. We are all dressed in the combat uniforms of our day: my olive drab, Dad's officer's jacket and cavalry boots, Bela's blue wool, Bill Bigbee's butternut grey, Peter Zachary and the Lombard Brothers in deerskin, Solomon's and Consider's homespun, Humphrey Tiffany in his black Puritan dress…and Levi sits alone in his old farmer's clothes, sweated, worn, a man with calloused hands.

For an instant no one speaks.

Even Peter Zachary with his cohorts, Humphrey and Consider Tiffany and the Lombards, sits silently, as if stunned by the weight of our gathering, the wars in which each of us fought now a palpable presence, the wars that made America, our memory of enemies, weapons, and killings.

Then Levi speaks.

You know what you all got in common?

You all fought in America's wars as citizen soldiers, the men who made this country into all it has become. Fought the Indians. The Revolution. Took the south and the southwest. Then the Civil War. World War One. World War Two.

William Green grunts.

Only the south lost.

Levi grins.

Not if you look at the United States today, it didn't. Look who's running the country. Texas. Mississippi. Their men come back to power after over a hundred years. Black folks still treated bad. You didn't lose, Bill Bigbee. You won. Jus' took a lot longer than you thought it would. That right, brother Bela?

Bela snorts.

Ahh, Levi, you always was a good talker!

Levi continues.

No, that's what each of you got in common. A war that made the country into all it has become. Your fight against the Indians, Humphrey, your conquest of the south, Peter Zachary, your rebellion in Texas, James Lombard, why they give us the land. Your Revolution, Solomon, it brought us freedom. And, Bela, the Civil War done saved the Union, freed the slaves. And, boy...he looks at me and my father...you and your Daddy, your wars made the country a world power.

But look what we done got out of all those wars.

All that greed while people, even little children, starve all over the world. Homeless people on our streets while others drive those big cars and drink that fine wine, live in those big houses, eat those six course meals.

All those awful weapons.

Black folks still second class citizens.

Is this what the death of men was about?

Solomon interrupts.

Like Rome!

And we was supposed to be great, the leaders of the world.

Levi

Like Rome!

All are silent for a moment then Peter Zachary suddenly snorts.

Like Rome?

Damn right we're like Rome. That was our aim from the founding of the nation. Look what we did. All the wars 'gainst the Injuns to take the land, taking the rest from Mexico, 'spandin' to the Caribbean, the Pacific, defending our rights in World War One 'n' Two, beating the USSR in the

Cold War, all these little wars in what you call Third World nations and what we called colored folks' countries. We, by God!, the good ol' US of A, owns the bloody world!

Whatcha think we got these weapons for?

He brandishes his Kentucky Rifle.

And Humphrey Tiffany, Consider Tiffany, the Lombard Brothers defiantly raise their weapons at his side

The United States done been attacked jus' 'cause it is like Rome. Got influence all over the earth. Hated 'cause of our power.

It's these weapons gonna save us.

Weapons protect us.

Weapons make us great.

Look at ya. All of ya used weapons in your life. Solomon you used that ol' Brown Bess. Bela, that .58 caliber musket. You, boy, that M–1 Garand and [he looks at my father] *you had that aeroplane. Now, we jus' got bigger weapons and, by God, we lucky that our leaders use 'em!*

He and the men at his side growl in agreement.

Levi Tiffany, white-lipped, shakes his head.

But what you got to know. Nobody ever gets what they really want out of a war. War wins the goals of the moment. But the meaning of those goals changes over time. The Indians you killed, Humphrey Tiffany and Peter Zachary, why they become our heroes. The freedom you so love, Solomon, why it evolves into power and greed and lust. The justice you so long for, Bela, why it becomes injustice. The defeat of the Nazis, boy, that victory made the nation great, gave us all this wealth … and nobody uses wealth that well.

Humphrey Tiffany, Consider Tiffany, Peter Zachary, the Lombard Brothers in chorus

We had to do it. Had to. Went with being a man.

Bela Tiffany and Solomon Lombard reluctantly join them

Sometimes you got to kill. We hated it. But sometimes a just cause requires killing.

William Bigbee Green

Only what's just today ain't just tomorrow.

My father

What makes this so important. We're the last...the last of that long, long line of citizen soldiers. Maybe our sons who fought in Korea and Vietnam join us. But, together, we are the last, the last of the line that volunteered for our war, fought, did our duty, then went back to civilian life.

They got professionals now. Men like us don't have to fight no more.

Bela

Or get kilt.

Me

Get wounded.

Bela derisively

We can't even talk to them, these new men. They pay people to die for them.

William Green contemptuously

They ain't got the guts to risk death.

Bela

The guts to kill.

Solomon

But they lost something mighty important. That pride. We, by God, we were there!

William Green

And that shame.

Bela

Shame? Yes, shame. You are right Bill Bigbee. What I saw in the Wilderness, moaning men and rotting corpses. Piles of legs and arms cut off in bloody charnel houses.

We must be shamed.

My father

But it's what makes us men.

Solomon

All we got. Some generations have hard duty.

For freedom.

Bela reflects.

For justice.

Do you think those boys today even know what it means to be a man? Not serving. Never facing the elephant?

Levi murmurs.

But ain't it that way with everyone now, all Americans? Don't have to give nothing up for their country?

Me

I think it's changed us, no hard duty left.

Solomon

Soft? Like Rome after the Republic?

Levi

Maybe winning a war is worse than losing it?

William Green sardonically

I don't know 'bout that.

Solomon

The funny thing. They ain't that many of us.

My father

Whatcha mean, Solomon?

Solomon

Men who really fought in all those wars. Men who did the hard duty.

How many families like ours? We been in 'bout every war the nation fought. Always as volunteers.

Bela

You're right. Never been that many of us.

Solomon

Whatcha reckon? A coupla million? Three or four at the most, men who saw the battle lines? Maybe 600,000 kilt? A million plus wounded?

Bela

"The death of men..."

Solomon

And nobody today understands what it was really like. Nobody!

Me

Movies make it a joke.

William Green

Books too.

Me

What did Siegfried Sasoon write? "War was a fiend who stopped our clocks though we met him grim and gay."

Bela

And that's why we can talk through time. All our clocks got stopped when we were young men.

William Green

Never got started again for some of us.

He kicks ashes over the fire.

Only it don't work.

He looks at Levi then at me.

The two of you are on the right track. All these centuries of war just don't work.

IT JUST DON'T WORK.

Levi

Enemies and weapons.

All are silent for a moment.

And Levi murmurs.

The air at summer dawn when the stars just begin to fade....

William Green with tears in his eyes

The tenderness of a woman's caress, her rising breasts....

And Bela whispers.

The beauty of my photographs, all graceful things....

William Green, after a long silence.

It's time to go, ain't it?

He reaches over and picks up his .58 caliber musket, holds it for a long time. Stares down at it then gently places it on the ground. He walks off into the fog of battle, alone, waving at us over his shoulders.

Peter Zachary shakes his head with contempt, stares after his grandson who, as he walks into the mist, has so purposely left his weapon behind him.

He yells after William Green, a great bellow of rage.

> *You are mad. I don't care if you are my grandson, you are mad. Giving up your weapons only leads to destruction. When the terrorists attacked us, look what we did: used those planes to bomb 'em. Sent our men in, those professionals you so decry, why, they saved us! Weapons are the only tools we have to keep madmen and criminals under control. Without 'em we are nothin'.*
>
> *Without your Brown Bess, Solomon, there woulda been no country. Without your old .58 musket, Bela, the slaves never woulda been freed. Without the weapons we carry, there woulda been no nation.*
>
> *Our weapons make the nation. Save our weapons! Use 'em!*
>
> *Killing at a distance, killing so many at one time, simply makes it better for us. Protects our men, kills more efficiently.*
>
> *We will not leave our weapons behind!*
>
> *Grandson,* [he yells into the fog of battle] *you done gone crazy.*

And he with Humphrey Tiffany, Consider Tiffany and the Lombard Brothers, stalk off into the fog of battle, carrying their weapons at port arms, ready to fire if needed.

We stand by ourselves, alone, Solomon, Bela, Levi, my father, and myself.

We stare at each other intently.

Solomon heaves a huge sigh, the slightest smile of regret upon his face as he tightly grips his Brown Bess.

> *My country right or wrong, the one I helped to make. When it's attacked by madmen, we must respond with the weapons of the day.*

He follows Peter Zachary into the smoke, carrying his Brown Bess.

Bela and Levi stare at each other for a long, long moment. Bela reluctantly shakes his head, tears in his old, brown eyes.

> *You know how I hate the killing, Levi. How I hate it! Always hated it. But sometimes you got to fight, Levi. Sometimes you just got to fight.*

He turns with an awkward wrench, almost runs as he, too, carries his weapon into the distant murmur of battle.

My father follows him, his .45 caliber pistol still strapped to his waist.

> *Remember all I taught you, son* [he calls over his shoulder]. *Remember all I taught you. Weapons to defend yourself against enemies, the heart of a man's soul.*

Levi and I stand alone.

Levi

> *I'll wait for you on the other side, boy. Do what you must do.*

William Green, in the distance, a voice barely heard

> *You're on the right track, boy. Wars done changed in your lifetime, those awful weapons that kill at a distance, soldiers now only professionals. Never meant to be that way. Great thing about America. Men in war were called up do their duty. Men don't give up nothin' for their country anymore. Don't know who they are as men. What being a man is all about. Give up sumpin' important, maybe your life, for your nation, your family, your town.*

And Levi, with him now, calls back.

> *Enemies and weapons, son, shame, what you have to teach the nation. With all those murderous guns in the world today, be a leader, help men see what they must do, how they must change, the way we all did in our times. Look at us, the way your Grandpa Solomon broke from his past to help found his country, Bela fought to free the slaves. You got to change like them. Got to show men it's these weapons they make to kill the enemies they so fear that will destroy the world.*

William Green murmurs in chorus with Levi.

> *Remember what Aeschylus inscribed on his tombstone:*
>
> *...his courage...the hallowed field of Marathon could tell....*
>
> *Courage, boy, courage, all a man has ever got to take him through his life.*
>
> *Courage, boy, the courage to do the right thing, no matter the cost, that's what being a man is all about.*
>
> *The courage you had when you crossed that bridge under fire, the courage we all we had when we faced our enemies on the battlefield, courage all a man ever has to face hard duty....*

Their chorus grows fainter, words lost in the mutter of artillery fire.

And I stare after them, these, my forefathers, the last of the line of citizen soldiers, as I struggle with what my nation has become, courage my inheritance as a man, courage all they gave me.

I shrug my shoulders then take my M–1 Garand, my weapon of World War Two, and lay it gently on the hard earth besides William Green's musket.

In the distance I hear the mournful mutter of shellfire as I follow my forefathers into the fog of war.

A Few Books on War and Peace, Violence and Compassion

These are books that, over this fifty-year search, have meant most to me. Some, unfortunately, are out of print but, taken together, they point toward a life beyond the weapons of our fathers.

Fiction

Bowen, Robert. *The Weight of the Cross*. New York: Alfred A. Knopf.

Camus, Albert. *The Plague.* New York: Alfred A. Knopf, 1948.

Conrad, Joseph. *The Heart of Darkness.* Included in *A Conrad Argosy,* New York: Doubleday, Doran & Company, 1942.

Dostoevsky, Fyodor. *The Brothers Karamazov.* New York: Grosset & Dunlap.

Hemingway, Ernest. *Farewell to Arms.* New York: Charles Scribner's Sons, 1929.

Jones, James. *From Here to Eternity*. New York: Charles Scribner's Sons, 1951.

Jones, James. *The Thin Red Line*. New York: Charles Scribner's Sons, 1962.

Jones, James. *Whistle.* New York: Delacorte Press, 1978.

Tolstoy, Leo. *War and Peace*. New York: Greenwich House, distributed by Crown Publishers, Inc.

Nonfiction

Aho, James A. *This Thing of Darkness*. Seattle, Washington: University of Washington Press, 1994.

Baron, Robert. *Soul of America*: *Volumes One and Two.* Golden, Colorado: Fulcrum Publishing, 1994.

Bellesiles, Michael A. *Arming America: The Origins of a National Gun Culture.* New York: Alfred A. Knopf, 2000.

Bok, Sissela. *Mayhem*. Reading, Massachusetts: Addison Wesley Press, 1998.

Boorstin, Daniel J. *The Americans*. New York, New York: Vintage Books, 1958.

Boyer, Paul. *By the Bomb's Early Light*. New York: Pantheon, 1985.

Carr, Caleb. *The Lessons of Terror*. New York: Random House, 2002.

Cooney, Robert, and Michalowski, Helen. *The Power of the People*. Culver City, California: Peace Press, 1977.

Diaz, Tom. *Making a Killing*. New York: The New Press, 1999.

Drinnon, Richard. *Facing West*. New York, London, and Scarborough, Ontario: A Meridian Book, 1980.

Ford, Roger. *The World's Great Rifles*. London: Brown Packaging Books, Ltd., 1998.

Friedman, George and Meredith. *The Future of War*. New York: Crown Publishers, 1996.

Green, Martin. *Tolstoy and Ghandi, Men of Peace.* New York: Basic Books, Inc., 1983.

Halberstam, David. *War in a Time of Peace*. New York: Scribner, 2001.

Hallie, Philip. *Lest Innocent Blood Be Shed*. New York: HarperCollins, 1979.

Hofstader, Richard, and Wallace, Michael. *American Violence*. New York: A Vintage Book, 1971.

Hurst, Jack. *Nathan Bedford Forrest*. New York: Alfred A. Knopf, 1993.

Jahoda, Gloria. *The Trail of Tears*. New York: Wing Books, 1975.

Keen, Sam. *Faces of the Enemy*. New York: HarperCollins, 1988.

The King James Edition of the *Bible.*

Kirschner, Allen and Linda. *Blessed Are the Peacemakers*. New York: Popular Library, 1971.

Kohn, Stephen M. *Jailed for Peace*. Westport, Connecticut: Greenwood Press, 1986.

Larson, Eric. *The Story of a Gun*. New York: Vintage Books, 1994.

Leopold, Aldo. *A Sand County Almanac*. New York: Sierra Club/Ballantine Books, 1966.

Lynd, Staughton. *Nonviolence in America: A Documentary History*. New York: Bobbs-Merrill, 1966.

McWhiney, Grady, and Jamieson, Perry D. *Attack and Die*. Tuscaloosa, Alabama: University of Alabama Press, 1982.

Merton, Thomas. *Ways of the Christian Mystics*. Boston, Massachusetts: Shambala Press, 1980.

Phillips, Kevin. *The Cousins' Wars*. New York: Basic Books, 1999.

Raggnhild-Von Hasse, Ragnhild, and Lehmkuhl, Ursula. *Enemy Images in American History*. Fiebig, Providence, Rhode Island: Berghanhan Books, 1997.

Rhodes, Richard. *The Making of the Atomic Bomb.* New York: Simon & Schuster, 1986.

———. *Dark Sun*. New York: Simon & Schuster, 1995.

Russell, Carl P. *Guns on the Early Frontiers*. Regents of the University of California: Bison Books, 1957.

Sagan, Eli. *The Honey and the Hemlock*. New York: Basic Books, 1991.

Schultz, Eric B., and Tougias, Michael J. *King Philip's War*. Woodstock, Vermont: The Countryman Press, 2000.

Sherry, Michael S. *In the Shadow of War.* New Haven and London: Yale University Press, 1995.

Slotkin, Richard. *Gunfighter Nation*. New York: Harper Perennial, 1993.

Todorov, Tzuetan. *Facing the Extreme*. New York: Metropolitan Books, 1996.

Weinberg, Arthur and Lila. *Instead of Violence*. Boston, Massachusetts: Beacon Press, 1963.

Williams, Michael. *They Walked with God*. Greenwich, Connecticut: Fawcett Publications, 1983.

Wink, Walter. *Engaging the Powers*. Minneapolis, Minnesota: Fortress Press, 1992.

Wood, Edward W. Jr., *On Being Wounded.* Golden, Colorado: Fulcrum Publishing, 1991.

Woodruff, Paul, and Wilmer, Harry A., eds. *Facing Evil*. Peru, Illinois: Open Court Publishing, 1988.

Zimring, Franklin E., and Hawkins, Gordon. *Crime Is Not the Problem: Lethal Violence in America*. New York and London: Oxford University Press, 1997.

Zinn, Howard. *A People's History of the United States,* New York: Harper and Row, 1980.

Zuckerman, Michael. *Peaceable Kingdoms: New England Towns in the Eighteenth Century*. New York: Alfred A. Knopf, 1970.

Poetry

Owen, Wilfred. *The Collected Poems of Wilfred Owen.* New York: A New Directions Paper Book.

Sassoon, Siegfried. *Collected Poems, 1908–1956.* London and Boston, Faber and Faber.

✯ ✯ ✯

Sources and Ways of Working

In addition to the books noted above, I depended heavily on the following for facts and dates of American history.

American National Biography. New York: Oxford University Press, 1999.

The Concise Columbia Encyclopedia. New York: Avon Books, 1983.

Dictionary of American Biography. New York: Charles Scribners Sons, and Collin MacMillan, London, Publishers.

The Encyclopedia of American Facts and Dates. Carruth, Gorton. New York: HarperCollins Publishers, Inc., 1997.

Statistical Abstract of the United States. Washington, D.C.: United States Department of Commerce, Census Bureau, 2000.

The World Almanac and Book of Facts. Mahwah, New Jersey: World Almanac Books, Imprint of Prime Media Reference, Inc., 2000.

It is from these books, those listed above in the bibliography, and an enormous range of other books that I constructed my Chronologies of Violence and Peace, my brief biographies of People of Peace, and found the quotations I used for those Americans over three hundred years of our history.

I thank the author of every book I read for helping me discover my American past and the men of my family who helped make the country.

The facts about my male forbears came from a variety of sources:

Humphrey Tiffany: *The Tiffany Family Genealogy.*

Consider Tiffany: *Ibid.*

Solomon Lombard: My father's Sons of the American Revolution application.

A brief biography written as part of the history of Harvard University.

Personal visits to Cape Cod and the probable site of his birthplace in Lombard Hollow, Truro, Massachusetts.

Peter Zachary: My mother's Daughters of the American Revolution application.

The Lombard Brothers, James and Stephen: *The Redditt Family Genealogy.*

William Bigbee Green: Family Bible.

Letter from State of Alabama re: his Confederate service.

Personal visit to his birth place and farm.

Bela Bentley Tiffany: *The Tiffany Family Genealogy.*

History of the Thirty-Sixth Regiment, Massachusetts Volunteers.

Levi Darwin Tiffany: *The Tiffany Family Genealogy.*

His Journals.

Edward W. Wood: Personal knowledge. Owner of all his papers.

These various family papers gave me the required knowledge of facts about my male ancestors who helped to make the nation. Their lives were liberally interpreted, acts of the imagination, in the choruses, based on years of reading histories about the way the nation was tamed from its first settlement in the 1600s until after the Civil War. I wanted to present the violent world these men must have experienced as they fought to take the land from the Native American, battled to dominate the environment, slavery normal to their lives.

Their world, the world of my male ancestors, was a hard and unforgiving place.

They had to be hard and unforgiving themselves in order to survive. Their pioneer world was not a time of mercy and compassion. Their inheritance from Europe, particularly those who fled Ireland and Scotland for the south, was one of great violence.

And, so, my interpretation of their lives is contained in the choruses, my understanding of their pioneer world, the struggles and battles they experienced.

Their feelings, their attitudes still help shape America, especially in our continued acceptance of the making of enemies and the improvement of weapons every generation.

It is those feelings and those attitudes I am certain that finally must be changed if we are ever to discover any peace in this very hard and cruel world.

Finally, I used these sources as the base for reporting on the numbers killed in American wars. Over my years of reading about American wars I discovered that there were so many different numbers for those killed and wounded in each war that the only sources I could accept were two above: *The Statistical Abstract* and *The World Almanac.* They supported and validated each other.

And, thus, the completion of this book, its contents and conclusions, rests on the shoulders of others. To them, to all of them who have written of the history of America, given me its facts and its stories, the people and events of that history, to all who have struggled with issues of war and peace, compassion and violence in America, I am eternally grateful.

Without their contribution I could not have written this book and the United States would not have its great potential.